The Strategic Project Office

Second Edition

PM Solutions Research

Editor
J. Kent Crawford
PM Solutions, Inc
Glen Mills, Pennsylvania

The Strategic Project Office, Second Edition
J. Kent Crawford and Jeannette Cabanis-Brewin

The Strategic Project Leader: Mastering Service-Based Project Leadership
Jack Ferraro

Project Management Maturity Model, Second Edition
J. Kent Crawford

Optimizing Human Capital with a Strategic Project Office: Select, Train, Measure, and Reward People for Organization Success
J. Kent Crawford and Jeannette Cabanis-Brewin

Effective Opportunity Management for Projects: Exploiting Positive Risk
David Hillson

Managing Multiple Projects: Planning, Scheduling, and Allocating Resources for Competitive Advantage
James S. Pennypacker and Lowell Dye

The Strategic Project Office: A Guide to Improving Organizational Performance, First Edition
J. Kent Crawford

The Superior Project Organization: Global Competency Standards and Best Practices
Frank Toney

The Superior Project Manager: Global Competency Standards and Best Practices
Frank Toney

Project Management Maturity Model: Providing a Proven Path to Project Management Excellence, First Edition
J. Kent Crawford

The Strategic Project Office

Second Edition

J. KENT CRAWFORD

with Jeannette Cabanis-Brewin

CRC Press is an imprint of the
Taylor & Francis Group, an **informa** business
AN AUERBACH BOOK

pmsolutions
»research

CRC Press
Taylor & Francis Group
6000 Broken Sound Parkway NW, Suite 300
Boca Raton, FL 33487-2742

Printed in the United States of America on acid-free paper
10 9 8 7 6 5 4 3 2

International Standard Book Number: 978-1-4398-3812-9 (Hardback)

Visit the Taylor & Francis Web site at
http://www.taylorandfrancis.com

and the CRC Press Web site at
http://www.crcpress.com

Contents

List of Figures

List of Tables

Foreword

Portfolio Strategy Group, Global Project Operations, Global Program Management Office, or Enterprise Project Management ... I have seen and used countless names for the organization that plays the central role in managing initiatives within a firm, agency, or line of business. These names hint at the promise and peril of such groups. Firms need to make their strategy "real"; in other words, they need to turn their strategy into operations.

Therefore, the ideal Strategic Project Management Office (PMO) ensures that the right projects are selected at the right time, done the right way, and deliver right and lasting results. Unfortunately, most PMOs never get even halfway to that goal. All of the research shows that most project management improvement efforts fail, most lasting only two years.

Why? My experience is that too many PMO leaders behave as if improved project management is an end in itself. Better templates, more skills, and a tool or two seem to be enough in many PMO directors' minds. Strategy and benefits management are thought of as someone else's job.

What then is the right way to build and sustain a strategic project office? Kent Crawford answers this question in the second edition of *The Strategic Project Office*. The first edition has been the foundation for much of my thinking about initiative management, and the second edition takes thought leadership to a new level.

Crawford leverages years of experience (I've been a client of PM Solutions for many years) to further improve an already excellent resource. Three elements of the second edition stand out:

- **Human Capital**. As implied above, too many PMOs are content with delivery of skill-based training. But, what about career paths? What about role descriptions? What about softer skills like holding a critical conversation or developing others? The second edition lays out a comprehensive approach to building a model PMO.
- **Strategy**. As Stephen Covey (influential author and management expert) says, "… building with the end in mind is a fundamental success habit." The second edition leverages recent research to reinforce the use of a Strategic PMO as a "strategy management center" to align corporate strategy with the project portfolio.
- **Collaboration**. Social media has the potential to revolutionize the way we collaborate on projects. I've seen it used to build methodologies, drive project communications, and brainstorm on requirements. To that end, the authors provide information on the use of collaborative and social media tools in project management.

The first edition of *The Strategic Project Office* sits on the top shelf of my bookcase. I'm sure that the second edition will have similar pride of place … when I haven't loaned it out to members of my team as homework.

Paul Ritchie, PMP
Director, Global PMO
Mead Johnson Nutrition, Inc.
Evansville, Indiana

Preface

What's New in the Second Edition?

If you were inspired to pick this book up off the shelf, chances are you already have some idea of the importance of good project management to today's organizations. Managing an endeavor with a fixed deadline, a unique product, and a budget cap is a very specialized art/science that has long been embraced by construction and the defense industries. More recently, project management was "discovered" by every industry where new products are the lifeblood of competition. In software development and other information technology/information system (IT/IS) application areas in particular, the value of sound project management has been underscored by the findings of research firms, such as the Standish Group, Gartner, Inc. and Forrester.[1]

However, sound project management of individual projects is no longer enough. While there still are some instances in which a company is almost entirely focused on one or two major projects at a time—small software development firms or capital construction firms come to mind—the reality for most businesses is dozens of projects throughout the company in various stages of completion (or, more commonly, of disarray). It wouldn't be at all uncommon for a company to have several new product development projects in process, along with a process reengineering effort, a quality initiative, a new marketing program, and a fledgling e-business unit. Widen the scope of your thought to take in facilities, logistics, manufacturing, and public relations and you begin to understand why most companies have no idea how many projects they have going at one time. And, when you consider that technology plays a role in almost all changes to organizations these days, and that technology projects have historically been challenged a hefty percentage of the time, the light begins to dawn. Unless all the projects that a company engages in are conceptualized, planned, executed, closed out, and archived in a systematic manner (i.e., using the proven methodologies of project management), it will be impossible for an organization to keep a handle on which activities add value and which merely drain resources.

As the saying goes, you can't manage what you can't measure, and unless all the projects on the table can be held up to the light and compared to each other, a company has no way of managing them strategically, no way of making intelligent resource allocation decisions, and no way of knowing what to delete and what to add. And the only way to have a global sense of how a company's projects are doing is to have some sort of project focus point: the Project Management Office (PMO).

Call it what you will—Center of Excellence, Project Support Office, Program Management Office, even Strategy Execution Office—a home base for project managers and project management is a must for organizations to move from doing a less-than-adequate job of managing projects on an individual basis to creating the organizational synergy around projects that adds value, dependably and repeatably. We believe that is why the PMO concept has taken off in such a startling growth curve in the past decade, necessitating a major update of this book.

Who Is This Book for?

The content of this book, in addition to supporting the coursework for our class ("The Ultimate PMO") is relevant to four principal segments of the project management community within an organization:

- Executive management project managers
- Project Office directors
- Managers of project managers
- Project managers

It is our hope that the book will make the rounds of all four groups because, without buy-in on all levels, the project office culture is very difficult to instill in an organization. Project managers need to understand the big picture of how their project fits into the organization as a whole, as a piece of the portfolio, and a part of the strategy. Managers of project managers and the PMO director require a game plan whereby they can improve project and project manager performance and take the lead in disseminating project knowledge throughout the company culture. Those who have already been tasked with the responsibility of spearheading the PMO will now have a handbook to guide them.

Executives need to be fully cognizant of the investment required of them in moving an organization to a higher level of functioning with regard to projects: they must champion the project philosophy, connect project performance to the overall bottom line and corporate strategy, and stand behind the implementation effort with the full power of their leadership. Because executive participation is so important and executives have so little time, you will find throughout the book important points for executives pulled out into Executive Tipsheets. Combined, these thumbnail summaries of the most crucial elements in each chapter provide the busy CEO with a crash course in the primary issues to be dealt with in each phase of implementing a PMO.

What's New in the Second Edition?

Back in 2000, Gartner, Inc. predicted that, through 2004, companies that failed to establish a project office *with appropriate governance* (author emphasis) would experience twice as many major project delays, overruns, and cancellations, as would those companies with a PMO in place.

At the time, this seemed like a bold statement; in fact, the PMO role that has emerged over the past decade is far more extensive than that. We will discuss the shape of the new "next-generation" PMO[2] in the Introduction and provide information on the changing status of the PMO from PM Solutions' 2008 *State of the PMO* research project, as well as from action research carried out at benchmarking events held from 2002 to 2009.

In Chapter 1, we provide new information on the Strategic Project Office (SPO) as a "strategy management center" to align corporate strategy with the project portfolio. This information has been derived from our *Strategy & Projects* research study, subsequent secondary research, and experience with clients in the field.

A new chapter (Chapter 6) on Project Portfolio Management (PPM) has been added, reflecting the vastly increased importance of PPM as part of the responsibilities and toolkit of the enterprise, or Strategic PMO. It showcases research on the maturity of PPM practices carried out in 2006.

Chapter 7 has been greatly augmented by materials we developed for our book, *Optimizing Human Capital with a Strategic Project Office* (Auerbach, 2005), covering the human resources (HR) aspects of managing project personnel.

In Chapter 8, we have added new material on an aspect of PMO infrastructure that scarcely existed at the time the first edition was published. The use of Web-based collaborative tools and social media tools in project management is growing and promises to revolutionize team communications and innovation practices.

Our chapters on PPM and knowledge management have been refreshed with expertise developed over the past decade in PMO and project management performance measurement. More than ever, the question of the value delivered by a PMO needs to be in the forefront of every practitioner's mind. Our value measurement framework has helped numerous companies to ascertain their organization's contribution to the bottom line.

Naturally, the entire book has been updated to the latest version of the Project Management Institute's standard, *A Guide to the Project Management Body of Knowledge (PMBOK®).*

Most noticeable is that the book is now structured in the same pattern as our Ultimate PMO (UPMO) class.* That means it covers four primary areas of knowledge and practice about the PMO:

- Governance and portfolio management
- Resource optimization
- Organizational change
- Performance measurement

* This PM College course is offered several times annually under the auspices of the Project Management Institute's Seminars World.

Table P.1 The UPMO Student's Guide to Using This Book

SPO SECOND EDITION CHAPTER	SECTION(S)/THEMES OF ULTIMATE PMO COURSE
*Introduction: The PMO Concept, Then and Now	PMO Purpose
*Chapter 1: The Strategic PMO: Aligning Programs, Projects, and Strategy	Purpose Functions
Chapter 2: Project Management Office Business Case, Organization Structure, and Functions	Function
Chapter 3: The Starting Gate: Assessing Your Current Condition	Strategy
Chapter 4: PMO Planning, Preparation, and Strategy	Strategy
Chapter 5: Establishing a Project Management Methodology	Governance
*Chapter 6: Portfolio Management	Governance/PPM Performance Management/ Benefits Realization
Chapter 7: Meet the Players: PMO Roles and Responsibilities	Resource Optimization
Chapter 8: The Technical Infrastructure: Using IT to Streamline Project Management across the Enterprise	Governance Resource Optimization Performance Measurement
Chapter 9: Changing Organizational Culture	Organizational Change Resource Optimization Performance Measurement
Chapter 10: Knowledge Management and the PMO	Governance Resource Optimization Performance Measurement/Benefits Realization

* New chapters.

And, it covers them in the order they are presented in the class, as shown above in Table P.1.

Finally, the appendices offer brief case studies of two stellar PMOs from our PMO of the Year competition slate of winners, from 2007 and 2008. These PMOs, chosen by an independent panel of expert judges, take the PMO concept to a new level. Many of them are actively implementing programs and strategies that we only dreamed about when the first edition of *The Strategic Project Office* was published.

In addition, replacing the paper-based appendices of useful templates and forms in the first edition, you now will find a CD-ROM of templates derived from PM Solutions' proprietary Project Management Community of Practice (PMCOP). The PMCOP materials have been refined in the fire, so to speak, of practical use by our consultants and clients.

How to Use This Book

First, read the introductory chapters for background on the PMO and its changing role in today's business. Then, do the quick organizational preassessment in Chapter 3 to find out what type of PMO is right for your organization. But don't set your sights too low. Our objective in writing this book is to help more organizations move to the most effective model of the PMO: the enterprise level or Strategic PMO.

We have seen companies fail in PMO implementations because executive leadership hands off the job of "creating a PMO" to one individual. But, the implementation of a PMO is less like a solo and more like a five-act production with a cast of thousands. The entire organization has to get involved. So, as Paul Ritchie suggests in the Foreword, pass this book around and use it as a tool to get both executives and PMO personnel "on the bandwagon."

We still agree with Tom Peters: "100 percent of everyone's time should be taken up by projects."[3]

But that's not enough. Just putting 50 ballerinas on stage and yelling, "Everybody dance!" does not make it *Swan Lake*, no matter how good their individual skills may be.

Everyone within a company and, to some extent, within that company's vendors, alliance partners, and other external stakeholders, needs to be working to the same beat. Management guru Stephen Covey is right: Interdependence is the name of the game, and the Strategic PMO concept is the way for people throughout your company to recognize and capitalize on their interdependencies, to manage and transfer project management knowledge, and to get into step with each other for the benefit of all.

Where do you start? Just turn the page.

Notes

1. James S. Pennypacker, *Troubled Projects: Project Failure or Project Recovery* (Center for Business Practices, 2006); also IT Projects: Getting Beyond CHAOS, PM Perspectives: http://pmperspectives.org/article.php?aid=22&view=full&sid=a9837d1ea09f04b17e13907d4789f044 (accessed October 27, 2008).
2. Margo Visitacion, "The Next-Generation PMO" (position paper, Forrester Research Group, November 2009).
3. Tom Peters, *Circle of Innovation* (New York: Alfred A. Knopf, 1997, p. 206).

Acknowledgments

This book may prove to be the exception to the old rule about writing by committee. As such, it's a great example of the efficacy of one of project management's signal features: the creative team. As with any product that distills the experience and knowledge of many people working together over time, it is a little hard to give all credit where it is due. I will do my best to name names here, with the uncomfortable feeling that someone is bound to be inadvertently left out. For any oversight, I apologize in advance.

Thanks first of all to Jeannette Cabanis-Brewin, PM Solutions' editor-in-chief, for polishing the rough core of my materials into a well-organized and readable book and guiding the manuscript through the publishing process with Taylor & Francis. Subject matter reviewers John Casey, PMP (PM Solutions), and Paul Lombard, PMP (PM College), were also instrumental as we refined the first edition and brought it up to date.

I offer a special "thank you" to my many students and clients. You have formed much of the knowledge base and case studies over the past many years. I so appreciate your contributed experience, insights, and wisdom to the collective mind of myself and PM Solutions. Through many shared challenges, in which we have agonized solutions, tested theories, and taken many risks, you have provided me with the insights I share in this book.

And, finally, I can't forget the many current and former associates of PM Solutions and PM College who, by their efforts on behalf of clients in the field, have helped to develop and refine our Strategic PMO processes, strategies, and knowledge. This team of experts created success after success in some of the most challenging cultures, industries, and environments. Yet, they continued to evolve, create, and design the most innovative and value-added project management approaches in the world. These many examples are embodied within the very structure you are about to explore. My deepest gratitude to the PM Solutions and PM College teams.

Introduction: The Project Management Office Concept, Then and Now

We've heard so much in the past few years about the PMO that it's hard to believe there was a time, not long ago, when the idea of an organizational center for project management was way out there on the fringe. But, project management wandered rootlessly throughout the organization for about a quarter-century before it became fashionable to build it a home in the Project Management Office (PMO). Let's examine how the management of individual projects evolved naturally into a project management center.

Early Days

Project management has been around for decades—some may argue for centuries. In the construction industry in particular, the idea of a work effort with a specific set of requirements and a deadline had been business as usual since the days of the pyramids. However, the concept of the project as an organizing principle and a management specialty—with its own techniques, tools, and vocabulary—had its beginnings in the twentieth-century military. Like many other features of post World War II America, the project and its supporting software and techniques were a spin-off of the first truly modern war effort.[1]

These military origins help to explain why the initial focus in projects was on planning and controlling. In fact, control might be considered the *raison d'etre* for project management: control of schedules, costs, and scope on endeavors that otherwise might careen over budget, over time, and/or fail to meet specifications.

That "out-of-control" feel that projects often inspire, however, has its roots in the way projects were superimposed on existing bureaucratic structures with their bulky communications mechanisms. Imagine the highly hierarchical command-and-control management of a military organization, and then imagine a short-term, deadline-sensitive effort with budget, personnel, and other resources drawn from multiple departments, divisions, and even branches of the service. Obeying the strictures inherent in the hierarchy while, at the same time, acting for the best interest of the project under such conditions would have been difficult at best, and was often downright impossible. But, this is the model we began from in project management. No wonder projects felt uncontrolled, mysterious; no wonder project management developed a reputation as both science and "art." It is interesting to note that the military was also the first to react to this situation, creating System Program Offices with semiautonomous program managers empowered to plan, execute, and complete projects (subject to "higher authority," such as the U.S. Congress). These System Program Offices were the precursors of today's project offices in the private sector.

An Evolving Structure

When project management's early tools—Gantt charts, network diagrams, PERT—began to be used in private industry, the new project managers faced a similar hurdle: business was also fashioned on the command-and-control model. Putting together an interdisciplinary team was a process fraught with bureaucratic roadblocks. The earliest uses of project management, in capital construction, civil engineering, and R&D, imposed the idea of the project schedule, project objectives, and project team on an existing organizational structure that was very rigid. Without a departmental home or a functional silo of its own, a project was often the organizational stepchild, even though it may have been, in terms of dollars or prestige, the most important thing going on. Thus was born the concept of the "matrix organization";

really a stopgap way of defining how projects were supposed to get done within an organizational structure unsuitable to project work. It was a "patch," to use a software development term, not a new version of the organization.

Today, those rigid, pyramid-shaped structures are changing shape. "Flattening" the organization means erasing the boundaries between functional silos. This trend is driven by both market imperatives and by the seamless communication made possible by modern communications technology. Multidisciplinary, team-based endeavors are now recognized as the only way to stay adaptive and flexible enough to succeed in a changing marketplace.[2] Many organizations that host projects now take a different tack; rather than forcing projects to fit within a bureaucratic structure, they embrace projects as an organizing principle.[3] Projects are no longer "something extra." they are the way work gets done at an increasing number of companies, from small start-ups to the likes of Hewlett Packard, IBM, USWest, Motorola, ABB, and many others. It's no accident that the majority of the applicants for our PMO of the Year Award are members of the Fortune 1000.

However, such change doesn't come easy. To take it out of the management context and put it in political terms, reorganizing a company's work around projects is the equivalent of moving from a feudal system to participatory democracy. Many of the participants in the PMO implementation courses I teach come from companies that have started "management by projects" initiatives in the past and failed—sometimes more than once.

Many times these failures are a result of the organization misjudging the magnitude of the change they were about to undergo. From many teaching engagements centered around implementing the PMO, my impression is that most of my students hold the misconception that a PMO is merely a project controls center that focuses on scheduling and reports. At one time, of course, this was true. In the old matrix organization, if a project was lucky to have a "project office," it was usually nothing more than a "war room" with some Gantt charts on the walls and perhaps a scheduler or two, people gifted with the ability to run the project management scheduling software of the day. This simple, single-project control office is what I call a Type 1 Project Office (see Chapter 2 for a full discussion of these levels).

A Type 2, or "department-level," PMO may provide support for individual projects, but its primary challenge is to integrate multiple projects of vrying sizes within a division (such as information technology (IT)), from small, short-term initiatives to multimonth or multiyear initiatives that require dozens of resources and complex integration of technologies. With a Type 2 PMO, an organization can, for the first time, integrate resources effectively because it's at the organizational level that resource control begins to play a much higher-value role in a project management system.

For an organization without any repeatable processes in place, such as the majority of software development organizations, which are at the first, or initial, level on the Software Engineering Institute's Capability Maturity Model,[4] these types of PMO organization are beneficial.

At the individual project level (Type 1), applying the discipline of project management creates significant value to the project because it begins to define basic processes that can later be applied to other projects within the organization.

At Type 2 and higher, the PMO not only focuses on project success, but also migrates processes to other projects and divisions, thus providing a much higher level of efficiency in managing resources across projects. A Type 2 PMO allows an organization to determine when resource shortages exist and to have enough information at their fingertips to make decisions on whether to hire or contract additional resources.

And, the Type 3 Strategic PMO applies processes, resource management, prioritization, and systems thinking across the entire organization.

The development of each of these types of project infrastructure provides a significant boost to process maturity. (For a fuller discussion of the relationship between PMOs and project management maturity, see Chapter 3.)

But, it's at Type 3—the enterprise or Strategic PMO—that the value-adding mechanisms of a PMO really reach warp speed. At the enterprise level, the PMO serves as a repository for the standards, processes, and methodologies that improve individual project performance in all divisions. It also serves to deconflict the competition for resources and to identify areas where there may be common resources that could be used across the enterprise. More important, an enterprise-level PMO allows the organization to manage its entire collection of projects as one or more interrelated portfolios. Executive management can get the big picture of all project activity across the enterprise from a central source: the PMO; project priority can be judged according to a standard set of criteria, and projects can at last fulfill their promise as agents of enterprise strategy. Gartner, Inc. has identified five key roles for a PMO,[5] all of which are most effectively carried out at Type 3:

- Methodology center: Developer, documenter, and repository of a standard methodology (a consistent set of tools and processes for projects).
- Resource evaluator: Based on experience from previous projects, the PMO can validate business assumptions about projects as to people, costs, and time; also a source of information on cross-functional project resource conflicts or synergies.
- Project planner: A competency center and library for previous project plans.
- Project management consulting center: Providing a seat of governing responsibility for project management; perhaps, staffing projects with project managers or deploying them as consultants.
- Project review and analysis center: A knowledge management center where information on project goals, budgets, progress, and history are stored, both during the project life cycle and after, in the form of lessons learned.

Recent research has shown that, in addition to all these roles and functions, the Strategic PMO has taken on responsibilities once reserved for the C-level, or set aside for the human resources (HR) department. The majority of enterprise-level PMOs now play a key role in recruiting, hiring, training, and developing their own personnel and in doing performance reviews. They also have taken ownership of the project portfolio management process, putting the PMO at the nexus of strategy and tasks. In addition, the pressure to show the value added by the various functions, from IT to HR to PM, has led the PMO to take the lead in refining benefits realization processes and implementing performance measurement frameworks.[6]

Thus, more than a place or a set of people, the PMO is "a shared competency" designed to integrate project management within an enterprise. A Type 3 PMO can promote enterprise competency in project analysis, design, management, and review. And, says Gartner, "... given the appropriate governance, it can improve communication, establish an enterprise standard for project management, and help reduce the disastrous effect of failed development projects on enterprise effectiveness and productivity."[7]

Although admittedly many companies today still struggle to implement even Type 1 Project Offices, the focus of this book is on the Type 3 Strategic PMO. Why? Because that's where organizations can get more bang for their buck—and realize organizational dreams at the same time. Like the matrix organization, lower-level project offices are a way station—a stage between the old-style organization and the new, project-based enterprise.

Why the PMO Matters *Now*

There's been a tremendous resurgence in interest in the discipline of project management in the past few years. The reason? Information technology, information services, and new product development organizations have "discovered" project management. A traditional part of the toolkit for construction and large government projects, project management now sparks interest wherever compressing time-to-market cycles is an issue; in other words, throughout the modern marketplace. As industries work hard to compress product lifecycles, to reduce costs, and to improve the quality of their deliverables, they are increasingly turning to project management.

Thus, the extensively practiced and researched disciplines of project control systems and schedule development have now come to find a home in less traditional areas, such as high-tech industries, where organizations are under increasing pressure to utilize product development funds more efficiently. There's been a shift of focus toward the business side of delivering high-tech products and services—a focus on the *process* and the *business* of managing projects.

With this microscope turned on the business side of IT projects comes the bad news. Most of them are not managed very well. In all fairness, project success rates in other industries may not be that great either, but they have not been subjected to the intense scrutiny that technology projects have been, for the simple reason that high-tech is one of the largest sectors in the economy, and, increasingly, it is difficult for any business not to rely on information technology as a crucial strategic resource. As IT moves out of the back office and into more mission-critical business processes like customer relationship management and e-commerce, the line between IT and other types of projects is blurring.

Failure: A Wake-Up Call

Most readers are probably familiar with the dismal technology project failure statistics that have been kept since 1994 by The Standish Group International Inc., a research firm in Dennis, Massachusetts. After a decade of improving results, troubled project rates rose in 2004, according to the Standish Group's CHAOS Report on software development projects. The 2004 report indicated that 71 percent of IT projects go awry.[8] While other studies found a lower percentage of troubled projects, it's certain that problems occur at an unacceptably high frequency. Research by PM Solutions' Center for Business Practices in 2005 indicated that 1,660 out of 3,952 projects performed by the surveyed organizations were troubled—an average of $30 million of projects at risk per organization.[9] Research sponsored by Computer Associates in 2007 noted that among United Kingdom-based IT organizations, a third of all projects implemented each year end up over budget due primarily to issues of interdependencies and conflicts between multiple projects coupled with the lack of control chief information officers (CIOs) have over project portfolios.[10] And research from the University of Oxford led by Christopher Sauer noted that a "volatility" in projects—frequent uncertainties of staff, sponsorship, deadlines, and scope—contributes to poor project performance.[11]

In construction, widely held to be the most mature industry in terms of project management, there was the loudly publicized failure of the Boston Central Artery/Third Harbor Tunnel (Big Dig) project in Boston, Massachusetts, which featured then-presidential candidate Sen. John McCain dressing down the project manager in front of the Senate Committee on Commerce, Science, and Transportation. At its inception, the project was expected to cost $2.6 billion, but a federal estimate in February 2000 put the actual price tag at $13.6 billion—a cost overrun of more than 500 percent.[12]

In the consulting field, such industry giants as Deloitte Consulting, PeopleSoft, and Andersen Consulting became targets of lawsuits in 1999 by companies furious that enterprise resources planning (ERP) and HR system implementations had dragged on for years, run millions over budget, saddled customers with incompetent consultants, and created a culture of dependency on the consulting firm.

Project failure, as Standish Group chairman Jim Johnson has noted, "... is everyone's problem."[13]

The solution? In 2000, Gartner, Inc. proposed that, as a Strategic Planning Assumption for companies, information system (IS) organizations that established enterprise standards for project management, *including a Project Office with suitable governance* (author emphasis), would experience half as many major project cost overruns, delays, and cancellations as those that fail to do so. They also noted that the IT software development project as presently managed is often 170 percent or 180 percent over budget.[14]

Why is this so important? Because time is money. If a project is late for an amount of time equal to 10 percent of the projected life of the project, it loses about 30 percent of its potential profits.[15] A study by McKinsey and Company has shown that high-tech products lose 33 percent of after-tax profits when they are late to market, but, lose only 4 percent when they are on time, even if they are 50 percent over budget.[16]

Failure: A Learning Experience

The good news in those bad statistics is that there is a trend toward improvement. In 1999, the Standish Group reported that project failure rates were falling.[17] Based on an examination of 23,000 software projects in companies of all sizes, in many industries since 1994, their research shows that project success rates are up, while cost and time overruns are down. In 1994, only 16 percent of application development projects met the criteria for success—on time, on budget, and with all the features originally specified. By 1998, 26 percent were successful. Large companies have made the most dramatic improvement. In 1994, the chance of a Fortune 500 company's project coming in on time and on budget was 9 percent; its average cost: $2.3 million. In 1998, that same project's chances of success had risen to 24 percent, while the average project cost fell to $1.2 million.

Johnson believes three factors explain these encouraging results: (1) a trend toward smaller projects that are more successful because they are less complex, (2) better project management, and (3) greater use of "standard infrastructures," such as those instituted through a PMO.

Although Standish's latest study still shows a high percentage of failed or challenged projects, with the improvement curve flattening, other researchers are beginning to question the methodology behind these numbers, pointing out that the definition of a failed project may actually mask successful and prudent business practices, such as knowing when to cancel a failing or redundant project. Such decisions can be the fruit of improved project portfolio management within the PMO.[18]

Nevertheless, one concrete benefit of this research has been the collection of an enormous amount of data on why projects fail.

Why Projects Fail *Infoweek* magazine put it succinctly in their August, 1996 issue: "The major cause of project failure is not the specifics of what went wrong, but rather the lack of procedures, methodology, and standards for managing the project." The project manager who is asked to manage a project with no methodology, no procedure, and no process to support them is going to be very challenged to keep that project under control. Some reasons for failure that are directly related to lack of a Project Office include:

- Project managers who lack enterprise-wide multiproject planning, control, and tracking tools often find it impossible to comprehend the system as a whole.[19]
- Ranges of acceptable project variances against key baselines are not established during project initiation or planning, thus, a kill or recover decision is not made early enough.[20]
- Poor project management/managers. Most of the reasons technology projects fail are management-related rather than technical. The old paradigm of promoting the best technical personnel to project manager level didn't work because technical ability is a poor indicator of project management ability, yet many enterprises have no processes in place to ensure that project managers are appropriately trained and evaluated.[21]

- There is a high correlation between lack of clear project sponsorship and failure. Executive support for/understanding of projects is lacking in many organizations.[22]
- Accurate project resource tracking is imperative to successful project management, but many organizations are hampered by awkward or antiquated time-tracking processes.[23]

What We Can Do About It Interestingly, many of the best practices for preventing failures are also directly related to PMOs:

- Enterprises that hold postimplementation reviews, harvest best practices and lessons learned, and identify reuse opportunities are laying the necessary groundwork for future successes.[24]
- A PMO shines as the repository for best practices in planning, estimating, risk assessment, scope containment, skills tracking, time and project reporting, maintaining and supporting methods and standards, and supporting the project manager.
- Sound project plans are realistic, up to date, and frequently reviewed; reviews focus not just on what has been done, but look forward to identifying risks and opportunities.
- Project metrics and milestones are defined, measured, and reported.[25]
- Experienced sponsors and project managers develop and maintain a "go/no-go" cancellation strategy. They don't hesitate to kill a project that becomes a liability, without indulging in blame and punishment.[26]
- Monitoring critical dates is imperative, and enterprise time-tracking software (usually Web-based for ease of use) has become a necessity for larger projects, multiproject environments, and dispersed project teams.[27]
- The project manager must be competent and experienced. Benefits of having a good project manager include reduced project expense, higher morale, and quicker time to market. The skills most executives cite as desirable in a project manager include: technology and business knowledge, negotiation, good communications (including writing ability), organization, diplomacy, and time management. Understanding the

business is more important than understanding technology. They must be able to define requirements; estimate resources and schedule their delivery, budget, and manage costs; motivate teams; resolve conflicts; negotiate external resources; manage contracts; assess and reduce risks; and adhere to a standard methodology and quality processes. Such project managers are not accidental, they are grown in an environment that trains, mentors, and rewards them based on performance in projects.

- Best-in-class enterprises have a process of due diligence to turn ideas into projects, using a standard checklist, addressing such issues as sponsorship, project plan, roles and responsibilities, and finance. Based on this checklist, a project is either given the go-ahead, further researched, or rejected.
- Projects should be carried out in a standard, published way, with a project method that sets planning and control standards, review points, the nature and frequency of project management meetings, and change control procedures. Project methods can be short and high level, but they must be clear and up-to-date.[28]
- Finally, projects must be aligned with the enterprise's strategic goals and vision, through the mechanism of project portfolio management, administered through the PMO.[29]

In a similar vein, as PMOs have spread, research and benchmarking on effective PMO practices and structure have proliferated with varying levels of accuracy and quality. By 2004, IT PMOs were beginning to take firm hold, but with mixed results; studies by *CIO Magazine* and Forrester Research showed that, while PMOs were being broadly implemented, the results were unclear.[30]

One study by APQC even seemed to suggest that project failure rates *increased* following the implementation of a PMO.[31] However, not all research is created equal, and often questionable assertions are made about project management based on sketchy data, primarily because many companies simply have not had the historical data about projects with which to create a baseline.

In Chapter 9, we will discuss some things to keep in mind when doing internal research and developing metrics by which to measure your PMO processes and progress.[32]

Tough Economic Times and the PMOs That Weather Them

By 2007, when we initiated our own broad survey of the status of PMOs, the burning question was not: "What is a PMO?" or "Why do we need one?" but "What kind of PMO do we need? and "How can we objectively measure the value it brings to the organization?" The results were astonishing. For example, in our 2000 *Value of Project Management* study, only 47 percent of the respondents had implemented a project office of any type. By 2006, 77 percent of the respondents to our *Project Management: The State of the Industry* survey had implemented PMOs; of those, 35 percent had an enterprise-level (or "strategic") PMO. In 2007, 54 percent of the respondents reported having an enterprise-level PMO in place. Even factoring in the differing research objectives of these studies, the upward trend is unmistakable, both in sheer numbers of PMOs and in the rising organizational clout.[33]

But, most important, those Strategic PMOs that had been in place for four years or longer seemed to making a definite difference in organizational performance. The results suggested that merely implementing a PMO is not a panacea. Instead, it is PMO *maturity* that makes a difference to the organization. As PMOs become more mature, our data suggests, organizational success metrics improve. In addition, the mature PMO takes on more roles: in portfolio management, in people management, and in performance management, further elevating its value to the organization.

In 2008 and 2009, the Strategic PMO faced unprecedented stresses: A global economic downturn left many companies reeling and scrambling for ways to cut costs. Yet our best-in-class PMOs continue to thrive because they allow companies to make the most of slim resources: streamlining the portfolio, accurately forecasting resource availability, and allowing changes in strategic focus necessitated by economic factors to be seamlessly carried out because the project portfolio management processes add nimbleness to the organization. And, the PMO received an unexpected boost from an unlikely quarter in 2009, when the U.S. federal government implemented the American Recovery and Reinvestment Act, along with unprecedented focus on excellent program management and transparency of results. Suddenly, federal agencies from NOAA to the USDA wanted Strategic PMOs. How this will play out over the long term is uncertain at this writing,

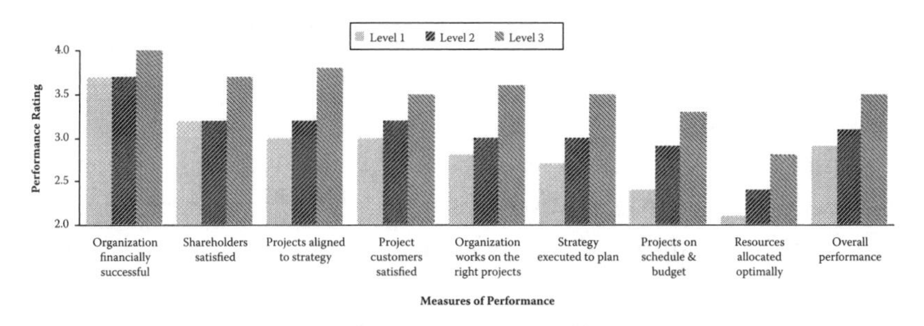

Figure I.1 Organizational performance by level of PMO maturity. Maturity is rated on a scale from Level 1 to 5 (immature, established, grown up, mature, best in class). "Performance improvement" is defined as rating higher on a scale of 1 to 5 on how well the organization performs in the eight measures of performance listed in the chart above (only Levels 1– 3 are listed because too few Level 4– 5 PMOs responded to draw accurate conclusions).

but, it is a positive note both for the status of project management and for the beleaguered taxpayer.

The Impact of the Mature Strategic PMO There is a strong correlation between organizational performance and the maturity of PMOs. Mature PMOs show significant improvement in organizational performance, as can be seen from Figure I.1.

Best of all, organizations with PMOs showed significant improvement *at each level* of PMO maturity; that is, for each incremental improvement in process maturity, there was a corresponding impact on organizational performance measures, including financial performance and customer satisfaction:

- 6.2 percent overall performance improvement from PMO Level 1 to Level 2.
- 14.6 percent overall performance improvement from PMO Level 2 to Level 3.
- 10.5 percent overall performance improvement from PMO Level 3 to Level 4.
- Organizations with PMOs at Level 3 maturity and higher showed a 16 percent budget/schedule performance improvement compared to those organizations with no PMO.

As PMOs mature, they are significantly better at meeting critical success factors, including having effective sponsorship, accountability, competent staff, quality leadership, and demonstrated value. They have significantly fewer challenges, including stakeholder acceptance, appropriate funding, demonstration of value, role clarification, conflicting authority, and consistent application of processes.

And (gratifying since we advocated this in our 2005 book on the HR aspects of managing by projects[34]) as PMOs mature, they are more likely to staff professional planners, schedulers, and controllers:

- Level 2 PMOs have 14 percent more planners, schedulers, and controllers than Level 1 PMOs.
- Level 3 PMOs have 24 percent more planners, schedulers, and controllers than Level 2 PMOs.
- Level 4 PMOs have 70 percent more planners, schedulers, and controllers than Level 3 PMOs.

Across the board, for respondents to this study (see an executive summary of the research in Appendix A), high-performing organizations are more likely to have an enterprise PMO (65.8 percent of high-performing organizations have enterprise PMOs compared to only 48.6 percent of low-performing organizations). The PMOs in high-performing organizations have been in place 29 percent longer (4.5 years) than in low-performing organizations (3.5 years); and high-performing companies have PMOs that perform a wider variety of functions, including strategy formulation, portfolio risk management, benefits realization analysis, contract preparation, outsourcing, project opportunity process development, resource assignment process development, management of a staff of project planners/controllers and business relationship managers, and resource identification and optimization.[35]

Wow! Given the obvious positive impact of these PMOs, why doesn't every company have one?

The Challenges of Implementing a PMO

Many times people will sign up for a PMO seminar thinking, "Just tell me how I can set up this administrative structure and I will go deploy it." They appear to believe Project Office is a clerical function, or that they can bring a small staff to bear to do administrative functions and *voila!* they have project management. Or, if their thinking is a bit more advanced, they perceive it as a project controls function—controlling cost, time, and resources within the individual projects. Unfortunately it isn't that simple because you are dealing with people, you are changing culture, building new processes, creating new approaches, integrating these elements across business units, and coordinating with teams of all sizes, technologies, complexities, and business interests. It's a worthwhile goal, but by no means a simple one to achieve.

The PMO is a function designed to facilitate the management of projects on one level, and improved management of the entire enterprise via project portfolio management and linking projects to corporate strategy. More than establishing an office and creating reports, it is infusing a cultural change throughout the organization.

Culture Change

It is a tremendous challenge to deploy and effectively apply these systems. Our work is cut out for us on so many fronts—both in system deployment and the educational arena—in order to get the best results from a Project Office. The complexity and magnitude of the effort of developing, designing, and deploying a full PO is too often underestimated. Let's look briefly at eight key areas of cultural change that the Project Office initiation will require. (For a full discussion of changing the corporate culture, see Chapter 9.)

Speed and Patience Years ago I studied and worked under Oliver Wight, the guru of Manufacturing Resources Planning (MRP) and MRP2. He had charted a number of MRP deployments, and he found that while you could never do one in less than 12 months, if you took much longer than 18 months, the failure rates dramatically increased. This pertains to the deployment of the project management culture throughout the organization; the longer the project duration, the greater the chance of failure.[36] Building a project management culture takes time. On the other hand, it is critical to meet clear objectives during deployment of the PMO or we will risk the possibility of a failed PMO project, with the participants losing sight of added value that project management practices can bring.

Therefore, the basic premise behind deploying a PMO is to move forward quickly, show results within six months, really begin changing the culture within the first year, and begin showing corporate results within a two-year time frame. But, be prepared that it will most likely take anywhere from two to five years to fully deploy a PMO.[37] (For a full discussion of how to structure the PMO rollout to show immediate benefits, see Chapter 4.)

Leadership from the Bottom Up Typically, what's happening in business today is that technology organizations are taking these studies that have been conducted by Gartner, Standish, and McKinsey very seriously. They see that their time-to-market is slow compared to either industry average or best-practice companies. They are finding that really the only way they can improve quality, improve time-to-market, decrease costs, improve timing, and improve the level of deliverables is to bring a new process to bear—something different from what they have used in the past.

IT processes and failures were thrown into the spotlight not just because of the research studies, but because of high-profile projects like Y2K and the Euro conversion. Thus, there is a tremendous amount of pressure on technology development projects to improve performance. They feel this pressure internally, but the failure data also has become a whipping stick for the other operational units to punish internal technology organizations with. Project performance has become a significant driver for people's careers and even for the existence of some internal IT departments. They must bring the organization to a position where it is actually delivering projects on time and within budget and with the quality that is desired by the customer.

So, unlike most organizational change projects of the past, we typically see the initiative to formalize project management begin on the department level, even on the project level. As technology efforts begin to show results, two things happen: (1) all the other the business units begin to come into the project teams as stakeholders of the organization, and (2) those business units see improved delivery performance on technology projects and ask themselves, "What can we do to improve our own performance? What are they doing right that we can adapt to our own projects?"

So, we are seeing a significant shift as IT brings project management to the organization. It's a grassroots change process quite different from anything traditional companies are used to.

A Systems-Thinking Perspective To effectively deploy project management throughout an organization, all the players must be on board. Everyone from the project team member on up to the executive sponsors of projects must understand what is happening with project management. This translates to an organizational setting in which virtually everyone who is touched by a project is impacted by what happens with the project management initiative. Ultimately this impact sweeps across the entire corporation. That's why effective PMOs are located at the corporate level, providing data on total corporate funding for projects, the resources utilized across all corporate projects, capital requirements for projects at the corporate level, materials impact, supplies impact, and the procurement chain impacts. To achieve corporate strategic goals, there will be strategic programs that generate strategic projects and those projects will of necessity reach across multiple divisions of the organization, and pull selected resources in to achieve that overall corporate objective. When corporate executives can effectively prioritize projects and make fact-based decisions about initiation, funding, and resources, they will be in a position to apply systems theory to their organization—to optimize the system (corporation) as a whole, rather than just tinkering with the parts (projects and departments). At this point, most corporations haven't yet achieved that level of sophistication.

Enterprisewide Systems Taking the need for common corporate data on resource projections as an example, we can see that all of the planning must be accomplished in a common database so that those resource projections can be summarized at the project level, then at the organizational level, on up to the corporate level, in order to understand the impacts of individual projects or new programs on the overall corporate resource pool.

For this to be possible, common systems must be established that integrate data and provide summarized integrated reporting in a timely fashion—not just with regard to resources, but also in the areas of capital funding, budgeted expenses, and the like.

At the organizational level, effective, integrated resources management, cost planning, and time tracking requires integration with corporate procurement systems, financial systems, time collection systems, and human resources systems. Systems integration at this level of complexity requires detailed specifications development and planning of its own accord.

Knowledge Management A whole new set of procedures and standards needs to be established along with a common mechanism for storing and sharing that information. Along with this goes the training process and data collection routine that must be established to get information into this database before knowledge transfer can take place.

One difficulty organizations face is that project management is a fairly new discipline in terms of the knowledge base and standards that exist to support practitioners. What standards exist can be found in *A Guide to the Project Management Body of Knowledge (PMBOK® Guide)*,[38] but these are limited primarily to the management of individual projects, not of an entire project-based organization. Few organizations have kept a history of lessons learned on projects done in-house, or possess standards for data collection of this kind. Therefore, most organizations are very new to this business of project management and are unable to rapidly develop this complex, integrated system that is necessary for accurate data collection and reporting.

Managers and project managers, program managers, will make good decisions with good data. Without good data, decisions are going to be very poor. So, the organization is faced with a very complex integrated system and process that they have very little knowledge how to deploy. That's why Gartner Group, Inc. recommends incorporating a contractor or consultant in the implementation strategy. It's necessary to get folks in who have actually done deployments in the past to make the probability of success much higher.[39]

Learning (and Learned) Project Organizations If your company has a system in place for educating, mentoring, and evaluating project personnel, you are in the minority. When I work with customers, I often ask who is currently engaged as a project manager, project leader, team member, project support staff, and other key positions. I get many positive responses. But, when I ask how many have college or university degree project management, only a few respond positively, and those individuals usually have a master's certificate in project management. The skill set and knowledge you need to effectively deploy a project management initiative rivals the knowledge set of an MBA in terms of complexity and integration. Yet, we ask project teams and project managers to effectively execute without having the requisite education or, in many cases, experience to deploy these very complex systems and processes. Learning has to take place enterprisewide for the PMO to be most effective.

Open Communication Communication—a sticky issue even within project teams—must now become free flowing, not just within but between projects and up and down the organizational levels. Why is this so important? Because 80 percent of what we call the "art" of project management is just communication and all the traits that good communicators display: trust, integrity, and honesty. We spend a great deal of time teaching the science of project management, such as how to develop project plans, how to do Gantt charts and work breakdown structures (WBSs), and estimating, and so forth, but these things are actually fairly straightforward. The real challenge comes in blending the art of project management into the science.

Therefore, through new channels of communication set up by the PMO, it will become possible for the entire organizational culture, from chief executives all the way through project teams, to communicate in a common language and work together to understand the issues surrounding how projects are faring and how the issues on one project affect other projects and, ultimately, the organization.

The Objective: Results and Fast

All this costs money, so, at the end of the day, it is absolutely essential that an organization is able to quantify the value that project management brings. What does success look like? How will you know when you have arrived? Dr J. Davidson Frame, PMP, of the University of Management and Technology in Washington, D.C., has performed research[40] that identifies the "traits of competence" exhibited by successful organizations:

- Top management understands project management basics.
- Activity-based costing systems are in place.
- Effective order processing systems are in place.

- Effective training programs are in place.
- Up-to-date tools are provided for staff.
- Clear project management systems and processes have been established.

In addition, a research study sponsored by PMI and the University of California at Berkeley identified the following organizational benefits:

- Improved coordination of intergroup activities
- Enhanced goal focus on the part of employees
- Elimination of redundant or duplicate functions
- Centralization of expertise
- A standardized management approach[41]

This topic is also discussed in Chapter 9, but the key question is how can the initiative show results fast enough to avoid top management loss of interest? There are two ways to demonstrate the immediate value of a PMO: through short-term initiatives and project mentoring.[42] The short-term initiatives provide solutions to immediate concerns and take care of issues surfaced by key stakeholders. These are items that can be implemented quickly while at the same time they take care of organizational top-priority concerns. Examples include: support for new projects and projects in need; an inventory of projects (new product development, information technology, business enhancements, etc.); summary reports and metrics; informal training lunches; project planning or project control workshops; and templates.

In conjunction with the short-term initiatives, project mentoring is an excellent way to provide immediate project management value to projects that are in the initial start-up phase or are in need of support, without waiting for the implementation of formal training programs or process roll-outs.

Notes

1. Francis Webster, "Setting the Stage for a New Profession," *PM Network* (April 1999).
2. Glenn M. Parker, *Handbook of Best Practices for Teams* (Amherst, MA: Human Resource Development Press, March 1996).
3. Paul Dinsmore, *Winning in Business through Enterprise Project Management* (Chanute, KS: AMACOM, 1998).

4. Software Engineering Institute, "Capability Maturity Model for Software Development" (Pittsburgh, PA: Carnegie Mellon University, 2009).
5. M. Light and T. Berg, "The Project Office: Teams, Processes and Tools," Gartner Strategic Analysis Report (Stamford, CT: Gartner, Inc. August 1, 2000).
6. James S. Pennypacker, *The State of the PMO, 2007–2008* (Glen Mills, PA: PM Solutions' Center for Business Practices, 2008).
7. Light and Berg, "The Project Office."
8. James Johnson, *My Life is Failure* (West Yarmouth, MA: Standish Group International, 2006).
9. James S. Pennypacker, *Troubled Projects: Project Failure or Project Recovery* (Glen Mills, PA: PM Solutions' Center for Business Practices, 2006).
10. Computer Associates, press release, September 26, 2007, http://blogs.zdnet.com/projectfailures/?p=413 (accessed April 7, 2008).
11. Christopher Sauer, A. Gemino, and B. Reich, "The Impact of Size and Volatility on IT Project Performance," *Communications of the ACM* 50, no. 11 (2007): 79–84.
12. The chequered history of this project is archived at http://www.massdot.state.ma.us/Highway/bigdig/bigdigmain.aspx.
13. Jim Johnson, "Turning CHAOS into SUCCESS," *Software* (December 1999).
14. Light and Berg, "The Project Office."
15. Preston Smith and Donald Reinertsen, *Developing Products in Half the Time* (New York: Van Nostrand Reinhold, 1991).
16. Brian Dumaine, "How Managers Can Succeed through Speed," *Fortune* (February 13, 1989). Also Charles House and Raymond L. Price, "The Return Map: Tracking Product Teams," *Harvard Business Review* (January–February 1991).
17. Johnson, "Turning CHAOS into SUCCESS."
18. Jim Johnson, "Getting Beyond CHAOS," PM Perspectives: http://pmperspectives.org/article.php?aid=22&view=full&sid=a9837d1ea09f04b17e13907d4789f044 (accessed October 27, 2008).
19. Lauren Gibbons Paul, "Turning Failure into Success: Maintain Momentum," *Network World* (November 22, 1999).
20. Richard W. Bailey II, "Six Steps to Project Recovery," *PM Network* (May 2000).
21. Paul, "Turning Failure into Success."
22. J. Roberts and J. Furlonger, "Successful IS Project Management" (Stamford, CT: Gartner, Inc. April 18, 2000).
23. C. Natale, "IT Project Management: Do Not Lose Track of Time" (Stamford, CT: Gartner, Inc. May 9, 2000).
24. Bailey, "Six Steps to Project Recovery."
25. Paul, "Turning Failure into Success."
26. Bailey, "Six Steps to Project Recovery."
27. Natale, "IT Project Management."
28. Roberts and Furlonger, "Successful IS Project Management."

29. Pennypacker, *The State of PMO*. Also James S. Pennypacker, *Project Management Maturity: A Benchmark of Current Best Practices* (Glen Mills, PA: PM Solutions' Center for Business Practices, 2006).
30. Project Management 2004, *CIO Insight:* http://www.cioinsight.com/article2/0,1397,1620739,00.asp (accessed June 2004); Thomas Hoffman, "Project Management Offices on the Rise, *Computerworld,* http://www.computerworld.com/managementtopics/management/project/story/0,10801,83159,00.html (accessed July 17, 2003).
31. Jim Lee, "Project Management Performance Metrics" (study of the American Productivity and Quality Council presented November 2003 at the Center for Business Practices Benchmarking Forum, Philadelphia, PA).
32. Jeannette Cabanis-Brewin, "Lies, Statistics, and the PMO: Some PMO Research Compares Apples with Oranges and Gets Fruit Salad," Developer.com: http://www.developer.com/mgmt/article.php/3399851/Lies-Statistics-and-the-PMO.htm (accessed August 26, 2004).
33. James S. Pennypacker, *The Value of Project Management, Practices* (Glen Mills, PA: PM Solutions' Center for Business Practices, 2000); James S. Pennypacker, *Project Management: The State of the Industry Practices* (Glen Mills, PA: PM Solutions' Center for Business Practices, 2006); Pennypacker, *The State of PMO*.
34. J. Kent Crawford and Jeannette Cabanis-Brewin, *Optimizing Human Capital with a Strategic Project Office* (New York: Auerbach Books, 2005).
35. Pennypacker, *The State of PMO*, p. 4.
36. Johnson,. Getting beyond chaos.
37. Dianne Bridges and Kent Crawford, "How to Start Up and Rollout a Project Office" (paper presented at the Proceedings of the PMI Annual Seminars and Symposium, Newtown Square, PA, 2000).
38. PMI Standards Committee, *A Guide to the Project Management Body of Knowledge,* 4th ed. (Newtown Square, PA: PMI, 2008).
39. Roberts and Furlonger, "Successful IS Project Management."
40. J. Davidson Frame, "Understanding the New Project Management" (paper presented to Project World, Washington, D.C., August 7, 1996).
41. C.W. Ibbs and Young-Hoon Kwak, "Benchmarking Project Management Organizations," *PM Network* (February 1998): 49–53.
42. Bridges and Crawford, "How to Start Up and Rollout a Project Office."

The Authors

J. Kent Crawford, PMP (project management professional), is founder and CEO of Project Management Solutions, Inc. (PM Solutions), a management consulting and training firm headquartered in Glen Mills, Pennsylvania. The company specializes in applying the project management discipline throughout organizations to improve enterprise business performance. Prior to establishing PM Solutions, Crawford served as president and chairman of the Project Management Institute (PMI®). Crawford is a recipient of the PMI Fellow Award and the award-winning author of *The Strategic Project Office: A Guide to Improving Organizational Performance* (for which he won the 2002 David I. Cleland Project Management Literature Award from PMI), *Project Management Maturity Model: Providing a Proven Path to Project Management Excellence*, and *Optimizing Human Capital with a Strategic Project Office.* His latest book, *Seven Steps to Strategy Execution,* provides the framework for organizations to execute and deliver corporate strategy through the use of Strategy Performance Management.

Jeannette Cabanis Brewin is editor-in-chief for PM Solutions.

1

THE STRATEGIC PMO

Aligning Projects and Strategy

Strategy is the organization's game plan. And although that plan does not precisely detail all future moves of the organization, it does provide a framework for managerial decisions. Without such a framework, it's easy for an organization—even a small one—to spin toward chaos. A business is presented with new opportunities and challenges each day; without some guiding point of reference to anchor present decision making to the future, various functions in the organization can wind up working at cross-purposes to one another.

A strategy reflects an organization's awareness of how, when, and where it should compete; against whom it should compete; and for what purposes it should compete; it is the integrated vision and direction of the organization as well as the *manner* in which it derives, articulates, communicates, and implements that vision and direction. Strategy answers the question of how a company will position itself against competition in the market over the long run to secure a sustainable competitive advantage.

Far from being some sort of "soft" fluff, strategic planning is where risk management is born.

Strategy has, alas, become something of a buzz word. Many companies claim to have strategy that is, in fact, nothing but a wish list of outcomes or a shopping list of tactics. Often, companies fail to distinguish between *operational effectiveness* and *strategy*. Targets for productivity, quality, sales, efficiency, or speed masquerade as strategies. These targets, while essential to superior performance, do not move an organization toward a strategic position in the marketplace. As a seminal article in the *Harvard Business Review* stated, "Operational effectiveness means performing similar activities better than rivals perform them.... In contrast, strategic positioning means performing different activities from rivals' or performing similar activities in different ways."[1]

Vague strategies cannot easily be translated into the measurable objectives or metrics so vital to achieving these kinds of stretch goals. Unclear corporate and business plans inhibit integration of objectives, activities, and strategies between corporate and business levels. Poor strategies, simply, result in poor execution plans. Those who have labored long in the project management trenches know very well how painful it can be to execute a project perfectly only to find out it is considered a failure because it does not meet a business need.

Therefore, as project management has gained in popularity, corporate executives have struggled to find a way to link strategic business objectives with the individual projects they have been asked to authorize. Too often, projects are chartered that have little or no connection to the corporate strategy formulated by top management. The reason for this is simply the lack of an organizational entity with responsibility to map strategy to projects, and to monitor projects and portfolios to ensure that they continue to address strategic initiatives, even as these initiatives change over time.

Enter the Strategic Project Management Office (SPMO). The SPMO not only provides all the services discussed in earlier chapters to individual projects and department-level project offices, it serves as the critical link between executive vision and the work of the enterprise. By providing a standard organizational methodology for planning, executing, staffing, prioritizing, and learning from all the projects that comprise today's organization, the SPMO gives organizational life a coherence that has long been lacking.

Let's explore just what an SPMO can do for your organization.

Overview

A PMO brings project management expertise to bear on any project-related problem or opportunity, wherever and whenever needed. This could include any of the PMO functions or services covered in the Introduction, or shown in Table 1.1.

The Strategic PMO goes beyond the traditional project management categories, with an expanded role that links strategic objectives to

Table 1.1 Common Enterprise-Level PMO Services

STRATEGIC SERVICES	OPERATIONAL SERVICES	HR-RELATED SERVICES
Strategic alignment	Methodology	Competency assessment
Governance	Project management	Project management training
Project portfolio management	Knowledge repository	Mentoring
Priority management	QA/auditing	Resource management (leveling and allocation)
Executive reporting	Processes support	Hiring and performance reviews

individual projects and portfolios. Several of these additional areas of project control and coordination are discussed elsewhere in this book:

- **Project Management Maturity.** As the owner of the project management process, the SPMO assesses project management maturity and takes action to improve the practice of project management across the organization (covered in Chapter 3).
- **Project Quality Management.** An enterprise-level standard and process for quality must be established. This usually falls with the Quality Assurance organization; however, if a separate QA organization does not exist, the Strategic Project Office (SPMO) is responsible for ensuring that project management process quality is maintained and that project managers take necessary action to ensure the quality of product and service deliverables to customers (covered in Chapter 5).
- **Project Office Steering Committee.** The Director of the SPMO should chair the steering committee that will select, prioritize, and terminate projects; make resource allocation decisions; and provide guidance to project managers (covered in Chapter 7).
- **Process and System Interfaces.** It is vital that various systems within the enterprise share information. The SPMO, working with the information technology (IT) section, takes the lead in this effort to integrate project management software with the accounting, human resources (HR), and other systems. While the specifics of this subject are outside the scope of this book, some PMO software issues are covered in Chapter 8.
- **Creation of a Project Culture.** The SPMO, working with the HR department, takes the lead in creating the project management culture so necessary for many of the advanced topics

covered in this book to be possible at the enterprise level (covered in Chapter 9).

- **Resource Management across Projects and Portfolios.** Perhaps the most difficult job of a project office is to ensure that resources are assigned to projects according to their position on the prioritized list. This can be done in a number of ways, from having a resource manager within the SPMO who takes requests from project managers and negotiates for resources with functional managers to forming a strong liaison with the HR department, which performs the same service for the project office. Regardless of the mechanism and procedure devised, the SPMO is responsible for ensuring that key projects are not delayed due to resource shortages. Note that it is not the role of the project office to lead project teams. That job belongs to the project manager. Software aspects of resource management are discussed in Chapter 8.

In this chapter, we will discuss those responsibilities and functions of the SPMO that specifically relate to its integrative and strategic role in the organization:

- Linking corporate strategy to programs and projects. The SPMO provides the organizational home for taking the strategy document produced by senior management and converting it into the projects that carry out that strategy.
- Portfolio management, including project selection and prioritization, manages the interdependencies between and among projects, which can only be seen from the perspective of an SPMO. (This topic area will be covered in detail in Chapter 6.)
- Finally, the SPMO has a critical role in project manager competency and professional development, which we look at in detail in Chapter 7.

As you can see from this list, the role of the SPMO is itself integrated with the roles and responsibilities of other staff organizations within the corporation. The relationship between the SPMO and both line and staff organizations within the corporation *must* be worked out as part of the change to a project culture (Chapter 9).

In short, the purpose or mission of a Strategic PMO should be:

- To ensure that the enterprise invests in the best set of projects and programs and realizes the most benefits possible from these investments.
- To provide an organizational focus on improving the management of projects, programs, and portfolios.
- To optimize the capability and use of scarce resources.
- To raise strategic issues to the highest levels of the organization to facilitate effective decision making.

The Link between Strategy and Projects

The SPMO has two primary missions: (1) to improve the organization's project management maturity (a process discussed in detail in Chapter 3) and (2) to "link the organization's projects to its strategic plans."[2]

The latter of these two—linking strategy to projects—remains revolutionary thinking in some organizations. On a consulting engagement to help an organization improve its project management practices, the author had been speaking to a group of senior managers about the connection between what they set as corporate strategy and what was happening "in the trenches" with their real projects. After about an hour, a group manager tentatively raised his hand and asked, "So, there's supposed to be a direct linkage between strategy and projects?"

The concept of having someone in the organization (other than the marketing department) look at the strategic objectives with respect to ongoing projects is still new in many organizations, or, if not new, then various initiatives have tried to align strategies and activities with limited success. The good news is that over the past decade, the number of organizations establishing Strategic PMOs to correct this problem has skyrocketed.[3]

Tying the corporate strategic objectives directly to the activities designed to achieve them plays directly into the development of project management maturity as well. It speaks to the need, identified by the Software Engineering Institute, to "institutionalize" good project management practices throughout an organization. If project management processes are not supported by the upper management of the organization, they will not be applied uniformly throughout.[4] Recognizing the SPMO as the organizational entity chartered to carry out this mission is one way that executive management can act to institutionalize best practices.

It is a source of constant amazement that corporations pay hundreds of thousands of dollars developing and deploying a project management methodology, then make its use discretionary. Human nature is such that most of us, given a choice to adopt a new process or continue with one that "works for us," will opt for continuing in our old ways. In Chapter 9, we will discuss how to overcome the natural resistance to change by stressing the advantages of adopting a new way of doing things.

Unfortunately, it is outside the scope of this book to discuss all the ways in which corporations can identify and establish strategic objectives because it is in this step that the seeds of successful projects are planted. Ideally, project activities should be at the bottom of a "waterfall" in which corporate strategy is expressed as a set of long-term and short-term goals, and each of these goals is operationalized as a project or program designed to carry out the strategic rationale. Furthermore, the setting of organizational priorities is expressed, on the project level, as a set of metrics by which executives can determine if the project activities, in fact, are moving the company toward the desired goal. These metrics, feeding into the project prioritization system and into the reward system for project team members, complete the feedback loop, making the organization coherent. Project team members are rewarded for behaviors that make projects serve the overarching goals; company strategy becomes everyone's business instead of the yearly intellectual playground of a top few.

By contrast, in the past, projects were insulated by many layers of management from the strategic rationale, with the result that work was undertaken on a departmental level that either failed to advance key strategic initiatives, or, in some cases, was actually detrimental. Project managers are familiar with the frustrating case of the project that is delivered on time, on budget, and to specifications, but which is a failure in the larger organizational sense because it is irrelevant to the corporate mission or to the competitive stance of the company. Linking strategy directly to the projects that are organized to carry it out eliminates this frustration, and saves a good deal of time and money as well. We believed this intuitively, but did not rely on intuition alone. A study completed in 2005 by PM Solutions' research arm[5] showed a strong linkage between aligning strategy with projects and excellent organizational performance.[6]

Strategy and Projects Research Study

Strategic planning becomes meaningless in the absence of a way to execute planned strategies. Organizations execute their strategies through the creation of "strategic initiatives," comprising portfolios of programs and projects, which become the vehicles for executing the organization's strategy.[7] To what extent does integrating corporate strategy with project portfolio management contribute to organizational success? To seek an answer to this question, which has significant importance for executives and project managers alike, the Center for Business Practices conducted a survey in November 2005, targeting a broad spectrum of organizations. Representatives of 87 leading companies responded. The results? Companies using identified "best practices" most consistently also had the highest rates of project success.

Why Align Projects with Strategy?

Nine out of ten corporate strategies devised on the executive level never come to fruition.[8] One reason for this is found in a survey conducted by the Society for Human Resource Management and the Balanced Scorecard Collaborative: 73 percent of polled organizations said they had a clearly articulated strategic direction, but only 44 percent of them communicated that strategy well to the employees who must implement it. These companies "are like a body whose brain is unable to tell it what to do."[9] Perhaps out of frustration with these failures, many companies are spending less time on strategy; research has shown that 60 percent don't link strategy and budgeting, and 85 percent of management teams spend less than one hour a month discussing strategy.[10]

Project management research has shown that most companies, far from having a coherent model for managing the projects as a "portfolio," have at best a vague idea how many projects they have in the pipeline, how much they will cost, how they will be staffed, or who is qualified to run them, making strategic planning an exercise in fantasy. Studies of the failure of customer relationship management systems confirm that lack of knowledge about one's own company is a primary reason for project failure. Companies who do not know their starting position build future corporate plans not on a solid foundation, but on shifting sand. Furthermore, their leadership often does not understand what is wrong, or how to distinguish what needs fixing.

A glance at the impact that the Balanced Scorecard has had upon businesses gives us some clues. The scorecard emphasizes the linkage of measurement to strategy. The tighter connection between the measurement system and strategy elevates the role for nonfinancial measures from an operational checklist to a comprehensive system for strategy implementation. For the first time, the details of the project portfolio (what the Balanced Scorecard creators call the "strategic initiatives") become important to a company's strategic thinkers.[11]

How Alignment Resolves Project Management Problems

Many studies have cited the lack of executive support as a key contributor to project failure. Project managers complain that their projects do not receive the resources they need. Projects completed "successfully" by project management standards (on time, on budget, to spec) have been considered failures because they did not address a business need. All of these issues are alleviated in a company that ties strategic planning to portfolio selection and project execution.

Strategy & Projects: Research Findings

The *Strategy & Projects* report was the initial product of a three-part research project. Part One was a review of management literature to develop a list of practices for aligning projects and corporate strategy. We first identified those practices that lead to high performance through a search of the literature on the integration of strategy execution, portfolio, program, project, and performance management.

This research revealed a set of best practices that we organized into a framework adapted from the McKinsey 7S framework.[12] The elements included:

1. Governance
2. Processes
3. Strategy Management
4. Project Portfolio Management
5. Program/Project Management
6. Structure
7. Information Technology
8. People
9. Culture

Best practices were defined under each process area, based on the management research reviewed. These practices were used to develop the questions in the survey. The goal of the survey was to learn whether or not organizations that exhibit these practices are, indeed, high performing, to confirm whether or not the practices identified are really "best practices," and to identify those practices that are most critical to the success of the organization. Participants rated their organizations on the frequency of their use of the best practices against a 7-point scale, where 1 = "not at all" and 7 = "to a great extent."* Members of the Center for Business Practices Research Network (senior practitioners with knowledge of their organizations' project management practices business results) were invited to participate in a Web-based survey. Of 87 respondents, 84 completed the survey in its entirety. We compared high-performing organizations, low-performing organizations, and all organizations, focusing on whether high-performing organizations exhibited the identified best practices more than the average of all organizations and whether or not low-performing organizations were below average in exhibiting these practices.

How "Performance" Was Defined Most of the measures used to ascertain which organizations performed well are familiar to all project

* These seven process areas were later refined into a framework for integrating project management principles with strategic management across the enterprise. The results of this framework development were published as Seven Steps to Strategy Execution (Center for Business Practices, 2007).

stakeholders. The survey asked not only about the success of project management by conformance to schedule, budget, requirements, and so forth, but also about the overall success of the organization. Practices related to the organizational value of project management, such as the rational allocation of project resources, the skillful selection and prioritization of projects, and the alignment of projects to business strategy, are being supported today by organizations increasing use of portfolio management systems and processes. As project management becomes more and more essential to the achievement of strategic organizational goals, these practices will gain in importance for all project stakeholders. The performance measures included:

- The organization's strategies are executed according to plan.
- The organization's shareholders are satisfied.
- The organization is financially successful.
- Projects are completed on schedule and on budget.
- Project customers are satisfied.
- Project resources are allocated optimally.
- Projects are aligned to the organization's business strategy.
- The organization works on the right projects.

Participants rated their organizations on the frequency of their achievement of these measures against a 7-point scale, where 1 = "not at all" and 7 = "to a great extent." Organizations termed "high-performing" in the results reported better-than-average performance in all areas measured. In particular, high-performing organizations are significantly better than average in allocating project resources optimally, followed by completing projects on schedule and on budget, and executing strategy according to plan. Low-performing organizations are significantly poorer than average in allocating project resources optimally, followed by completing projects on schedule and on budget, and satisfying the organization's shareholders. These results are displayed in Figure 1.1.[13]

Key Findings A significant finding was that the results confirmed the best practices proposed by the management literature. This underscores the value of using the best practices outlined in the *Strategy & Projects* in executing an organization's strategy. High-performing

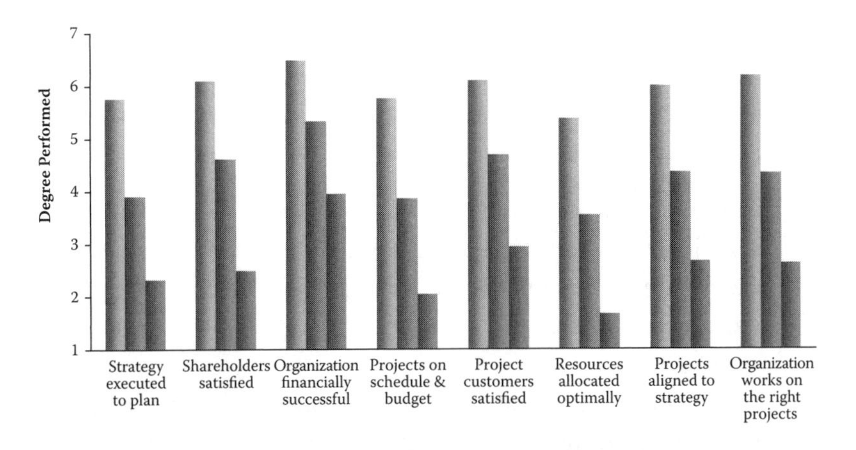

Figure 1.1 Performance Indicators. The left bar indicates the frequency with which each measure was reported by the top-performing companies in the survey; the right bar indicates frequency with which that measure reported the low performers in the survey. The middle bar expresses the mean of all companies.

organizations use best practices in all areas more than other organizations, consistently and significantly. Low-performing organizations consistently underutilize the best practices in all areas.

The Strategy & Projects Framework

Governance is the policy framework within which an organization's leaders make strategic decisions. With an effective governance framework all strategic decisions throughout the organization are made in the same manner. Each level within the organization must apply the same principles of setting objectives, providing and getting direction, and providing and evaluating performance measures. Using a common governance framework ensures that decisions are made the same way up and down the organization.

Perhaps not surprisingly, the most often used governance practice by high-performing organizations in the study is *having a well-defined strategy*. Let's examine the strategy management process, and look at the best practices identified.

Strategy Management Strategy management moves the organization from its present position to a future strategic position in order to exploit new products and markets. Strategy management is accomplished through the application and integration of strategy management processes, such as mission-vision formulation, strategy formulation, planning, execution, and monitoring/control. Best practices identified for strategy management include:

- Strategy performance is measured, compared to objectives, and activities are redirected or objectives changed where necessary.
- There is an understanding of the impact of projects or project management activities on the creation and implementation of strategy.
- The organization's strategic plans cascade down from corporate strategy to business unit strategy to portfolio, program, and project strategy.
- Corporate and business units assemble a strategic portfolio of programs and projects, and measure the strategic contribution of a program or project and adopt or reject programs/projects based on this information.
- As strategy cascades down the organization, performance measures are established at each level (business unit, portfolio, program, project) to link up with the strategic performance expectations of the entire company.

The most often used practice by high-performing organizations is having strategic plans that cascade down from corporate strategy to business unit strategy to portfolio, program, and project strategy; conversely, using project and program performance feedback to manage strategy execution is also a best practice engaged in by high-performing companies. The use of these practices also makes the difference between high performance on the enterprise level and just getting by. This is demonstrated in Figure 1.2.

And, the Best-Practice Structure? Corporate strategy affects the choice of organizational structures; likewise, organizational structures are important to the execution of corporate strategy. To execute strategy effectively, managers must make sound decisions about structures and develop methods or processes to achieve the needed integration of structural units. Organizational structures take many forms, each affecting the speed at which change can be brought about. They include line and staff structures, functionalized structures, matrix structures,

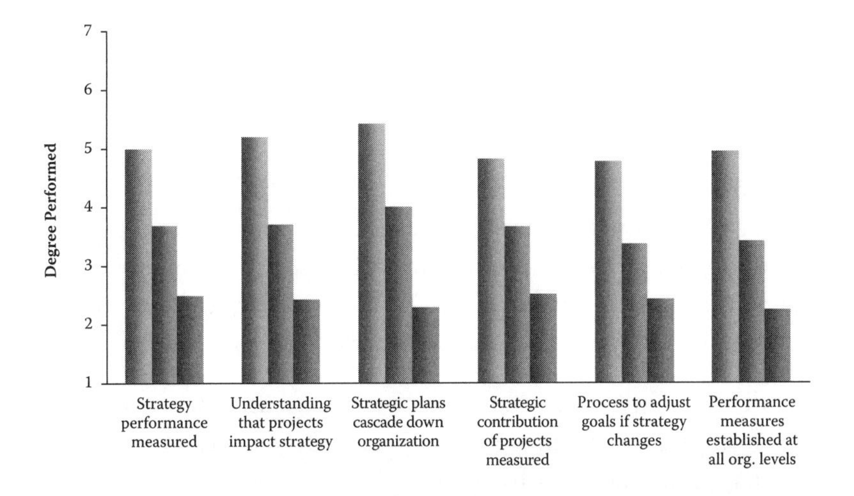

Figure 1.2 Projects and Strategy Alignment Practices and Organizational Performance. The left bar in each set indicates the frequency with which that metric was reported by the top performers in the survey; the right bar indicates the reported use of that best practice by low performers. The middle bar expresses the mean. Note the dramatic difference between high and low performers on each measure.

multidimensional matrix structures, strategic business units, laissez-faire structures, and virtual structures (listed here in order of their increasing ability to adapt to rapid changes in strategic direction demanded by changing market conditions). The best practices identified include:

- A strategic (enterprise) project office (sometimes called the Office of Strategy Management) plays a role in linking the organization's projects to its strategic plans.
- The company has an organizational structure (strategic project office, office of strategy management, strategic steering committee, etc.) that is responsible for managing strategy execution.
- Project management is clearly established and embedded within the organization's business management structure.
- Information about strategy and projects flows freely between business units facilitating strategy execution.

The most often used practice by high-performing organizations is *having project management clearly established and embedded within the organization's business management structure*, and, for most companies, this naturally translates into a Strategic PMO. The SPMO also plays a key role in another best practice: making a focus on strategy execution an important part of the organization's culture.

Top 10 Best Practices that Set High Performers Apart When it comes to aligning projects with strategy, best practices were used significantly more often by high-performing organizations than other organizations. Information technology best practices, in particular, set high performers apart. The practices are listed in order of their significance.

- IT tools integrate strategy execution management, portfolio management, program/project management, and performance management functions.
- IT tools are used to develop alternative strategic and project portfolio scenarios.
- Project management is clearly established and embedded within the organization's business management structure.
- IT tools provide information on the availability of resources.
- Senior management consistently rewards successful project behaviors.
- The enterprise project office allows the organization to manage its entire collection of projects as one or more interrelated portfolios.
- Program/project performance feedback is used for managing strategy execution.
- IT tools provide the capability to monitor and control risks, issues, and financials across portfolios.
- Project management is valued throughout the organization.
- The company has an organizational structure (strategic project office, office of strategy management, strategic steering committee, etc.) that is responsible for managing strategy execution.

As we continued to follow up on the findings summarized above, one striking, though admittedly, anecdotal correlation kept popping up: Nearly every organization in the top 20 performers is the recipient of at least one and, in some cases, many awards specific to their field of endeavor. Coincidence or proof of the power of aligning projects with strategy?

Best-Practice Examples

The Strategic PMO was a relatively new phenomenon in organizations when the first edition of this book appeared; few best-practice examples were available to use as templates. At the time, we predicted that the development of the SPMO as a feature of the modern organization would be one of the most exciting trends in organizational development this decade. However, we were unprepared for the speed and enthusiasm with which companies would adopt the SPMO framework and begin optimizing it for their industries and corporate environments.

In 2006, impressed by the way the PMO's influence was expanding throughout organizations, we initiated a competition to honor Strategic PMOs that were displaying not only our recommended practices, but often going them one better. The PMO of the Year competition, now in its fifth year, has brought public attention to PMOs that perform critical, strategic functions: PMOs with names like "Office of Strategy and Planning," PMOs with directors who are at the vice president level or even function as Chief Project Officers. In Appendix B, short case histories of the award winners from 2007 and 2008 are showcased.

Notes

1. Michael S. Porter, "What Is Strategy?" *Harvard Business Review* (November/December 1996): 61–78.
2. Nolan Eidsmoe, "The Strategic Program Management Office," *PM Network* (December 2000).
3. James S. Pennypacker, *State of the PMO 2007–2008* (Glen Mills, PA: PM Solutions' Center for Business Practices, 2008).
4. Software Engineering Institute (SEI), *Capability Maturity Model for Software* (Pittsburgh, PA: Carnegie Mellon University, 1993).
5. Jeannette Cabanis-Brewin and James S. Pennypacker, "Best Practices for Aligning Projects to Corporate Strategy" (paper presented at the Proceedings of the 2006 Global Congress, Anaheim, CA, Project Management Institute, 2006).
6. Center for Business Practices, *Strategy & Projects: A Benchmark of Current Best Practices* (Glen Mills, PA: PM Solutions, 2005).
7. Robert S. Kaplan and David P. Norton, *The Strategy-Focused Organization* (Cambridge, MA: Harvard Business School Press, 2001).
8. Jeannette Cabanis-Brewin, "PM Metrics: Moving toward a Balanced Approach," *Project Management Best Practices Report* (September 2000).

9. J. Mullich, "Human Resources' Goals Work Best When They're Tied to Company Success," *Workforce Management* (December 2003).
10. J. Hope and R. Fraser, "Figures of Hate: Traditional Budgets Hold Companies Back, Restrict Staff Creativity and Prevent Them from Responding to Customers," *Financial Management* (February 2001).
11. Kaplan and Norton, *The Strategy-Focused Organization*.
12. BuldingBrands.com. (2006) "The McKinsey 7S Model": http://www.buildingbrands.com/didyouknow/14_7s_mckinsey_model.php (accessed July 28, 2006).
13. Center for Business Practices, *Strategy & Projects*.

2

PMO Business Case, Organization Structure, and Functions

Where does "strategic project management" come from? How does it develop? Is it a business practice that one can purchase or study in school? Unfortunately, the term has been loosely applied to everything from multiproject software to courses on how to start a Type 1 project office. Our definition of strategic project management is that it is *an evolved practice*, closely paralleling the evolution of a business. We can define three stages in this evolution: (1) individual project management, (2) division or department-level project management, and (3) enterprise/corporate-level (which we will call *strategic*) project management. These three stages correlate to the types of project office (Figure 2.1) briefly described in the Introduction.

The Evolving Enterprise

Most businesses begin rather humbly. Even multinational corporations were once small concerns, perhaps even sole proprietorships. But, like any organism, as a business grows it gains complexity and its needs change. As the organization grows in complexity, projects begin to interrelate and have impacts on each other in more subtle ways; ways that might not be understood until each project is well under way and suddenly a conflict or dependency emerges, usually an unpleasant surprise.

Below is an imaginary case study that describes how that evolution normally takes place.

Case. In our example, Joe Merchant and his wife Jane decide to open a grocery business in Chinook, a small town in Wisconsin. They begin in a small storefront on Main St. and the locals, appreciating the prices and fresh produce, give them a good start in business.

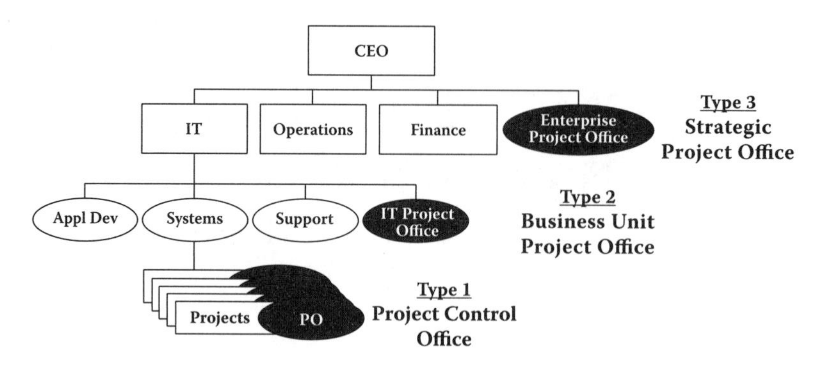

Figure 2.1 Three types of a PMO.

Analysis. Does Merchant Grocery Co. need project management? Probably not. This is purely a process-oriented business; they are involved in the same procedures day after day. However, the project of refitting the storefront and preparing for opening day could have been managed as a project, as could their initial marketing campaign.

Case. In two years Joe and Jane are ready to branch out and decide to start a second store in Walleye, a neighboring town. They purchase a new building, renovate it, stock it, hire and train clerks, and go through a start-up process much as they did in Chinook.

Analysis. Although Merchant Grocery doesn't realize it, it is using project management to make this change. The new store startup is a temporary endeavor, and they are creating a unique product (the new store). Probably they have artifacts common to project managers, derived from the lessons learned in opening the first store, for example, a checklist.

Case. The scene now jumps ahead 15 years. Joe and Jane have been very successful with Merchant Grocery. After opening their eighth store, they decided to centralize much of the operation by opening an office and warehouse facility in northern Wisconsin. The checkout lanes in all their stores are now served by computer registers and all stores are serviced by an automated inventory system. The 10-person IT staff in the central office writes and maintains the shipping/receiving system, the checkout system, and other applications.

Analysis. As astute business people, Merchant Grocery readily understands the need for control of business functions. In particular, they understand that, whenever changes are made around the company (whether it's building the next store or changing the shipping system), it is imperative to manage the project. At this point, the Merchant's concept of project management is on a *project-by-project* basis. Projects, as they come up, are executed by individual project managers (who may not even think of themselves as such) using whatever tools they have at hand and with which they are familiar. Some projects are managed by the PIN method (Post-It Note®), while others may be more formally managed using concepts from the *Guide to the Project Management Book of Knowledge* (*PMBOK®*). But, each project functions as its own entity. Changes within one project are not considered in another unless some information gets exchanged by the water cooler.

Case. Merchant SuperStores, as the company has recently been renamed, is now a robust organization with 15 stores all over Wisconsin and northern Illinois. Joe, as president, has initiated several projects, one of which is to build a new regional distribution center, along with a new warehousing system. Another project will entirely replace the shipping/receiving system in the existing warehouses. The shipping/receiving project is in final testing before it's discovered that if the project team had accelerated the project schedule by only four months, the new system could have easily been integrated with the new warehousing system. Unfortunately this lack of coordination will call for a rewrite of the shipping system once the warehousing system is implemented—costing the company over $100,000.

Analysis. As with most organizations that increase in complexity, Merchant SuperStores has gotten to the point where one project impacts another. At this point, the organization must insist that there be some modicum of control outside each project. Generally, at this point, an organization directs its attention toward a project management software tool. The implementation of project management software helps to ensure that, at least, those projects that are managed within the tool do not conflict with each other in important ways. However, there are still projects in the organization, say, in Merchandising, where Jane Merchant, the vice president, is launching a new program for getting customer feedback, that fall outside the purview of the software. Thus, project management is at a divisional or organizational level ... but not yet extending across the enterprise. At this stage, the organization (or not infrequently, a Project Office) will check schedules of projects against each other. At some point, utilization of key resources (such as database analysts) between multiple projects will become enough of a consideration (read: "problem") that the Project Office will monitor their usage, arbitrating conflicts when they arise. At this stage, many of the project managers will be keeping Issue Logs, instituting Change Management processes, etc., but there will be little consistency, and no control of these things by the Project Office.

Case. Merchant SuperStores now has 52 stores across a five-state swath of the Midwest and four automated distribution centers in strategic locations. The IT staff has a head count of 273, including the database staff, analysts, programmers, managers, etc. They now complete some 150 projects annually and have 15 full-time project managers on staff. Although these project managers report to the line of business IT teams, they are matrixed into the Project Office for expert support.

The Project Office has found it absolutely necessary to examine each project closely at initiation, sitting in on review of the stated scope and objectives with what used to be called the senior management oversight team (now renamed the Project Office Steering Committee at Merchant). There is now a standard PM software package in use within IT as well as other departments within the company, such as the Facilities Department. For this reason, the Project Office is now providing training on PM software internally.

Often, key resources are demanded by multiple projects. In these cases the Project Office steps in to arbitrate the demand. This is happening more and more now that so many projects are ongoing at the same time. Another problem that has come to the surface is that each time a resource moves from one project team to another, there is a learning curve because each project manager has his/her own ideas of what needs to be documented, how the documentation should be structured, etc.

Because Merchant IT has at least 15 projects in process at any given time, a company business unit or support department may be interacting with multiple project teams simultaneously. For example, the Purchasing manager is actually working with three teams right now.

She is very confused, and frankly, angry. One of the project managers asks her for sign-off on almost everything (design documents, scope statements, test results), while another (we'll call her Sue) pretty much "wings it;" she talked to the manager for a while, and came back two weeks later with a system of screens put together on the network, which they worked together to change. The third is somewhere in the middle. He goes through a rather formal review of certain major documents and asks for sign-offs at the end of what he calls "project phases," and then seems to go away for weeks at a time.

Analysis. It should come as no surprise that the things the Project Office at Merchant SuperStores finds itself needing to coordinate and manage fall into the nine Project Management Knowledge Areas (described in Chapter 3). Merchant has found that there are needs in each of the knowledge areas for higher-level coordination and management. They need a PMO with the reach and capability to handle all the following issues:

- PM software
- Schedules
- Resource coordination
- System interfaces
- Quality
- Change control
- Human resources
- Communications
- PM methodology
- Procurement
- Training
- Cost
- Risk

As the tale of Merchant SuperStores illustrates, increasing levels of sophistication in a business call for increasing levels of integration in project processes, much the same way that more complex businesses require better financial tracking. Obviously, more work is needed at Merchant to standardize the project management process and publish a methodology. Just as a financial system rolls up expenditures and revenues across many departments into a single budget, in a strategic project management system, individual projects (Type 1) roll up into programs, programs roll up into business unit project control (Type 2), and then, ultimately, those programs, which in some cases may stretch across the organization, roll up into enterprisewide, or strategic, project management. And, at that point, it is actually integrating a project management system that reaches across the entire enterprise, affecting all projects.

Ideally, the fully evolved Project Management Office (PMO) operating at the enterprise level will be capable of addressing all of the issues in the bulleted list in our Merchant Superstores example. Let's take a closer look at what's involved with each of those issues:

- **PM Software**. Assessing the need for project management software, conducting research to determine the best package or suite for the organization, going through the procurement process for that software, implementing the software in the organization, training those who will use the software, coordination and administration of the software when it is in operation, and support of the people using the software after implementation.
- **Schedules**. The estimation for, creation of, and (most important in a complex organization) coordination of schedules for the various projects. At the enterprise level, it is critical to determine and track the interrelationships of tasks from one project to another.
- **Resource Coordination**. The enterprise-level coordination of resources working on projects, most specifically of key resources, but to some extent of all resources. Usually these will be human resources, but not always. Key resources on a construction project may be construction materials, supplies, equipment, trucks, or pieces of difficult-to-acquire equipment, for example. Key resources in event management may be conference rooms in a hotel.
- **System Interfaces**. As the organization takes a higher-level (strategic) view of projects, it will become more and more important to achieve a tight interface between project management software and other administrative systems, such as:
    - *Financial Systems*. Adjusting the accounting/finance system to support project management in addition to the traditional support of functional departments; project estimates should integrate into departmental budgets; actual hours worked and other direct costs associated with project tasks should be downloaded to the project management software, enabling project managers to track variances; project material, supplies, and equipment costs must be allocated to the costs of the project.

    - *Procurement Systems.* Many projects involve outsourcing of hardware, software, and services. Procurement lead time for project-related materials, equipment, and other critical items must be reported to the project manager to avoid potential negative impact on the critical path.
    - *Human Resource Systems.* A project may require resources from many areas of the company and may have a certain skill set requirement for each. A skills inventory for all people in the organization should be established in the HR system and made available to the project manager through the project management software. Project schedules must take into account planned vacations, company holidays, and other time not available for project work. A good HR system will have that information readily available. If the project management software is to keep records on actual project expenditures, it must know what each person is paid, and that information is also available in the HR system.

- **Quality**. As projects are performed within multiple departments in the organization, it becomes imperative that an enterprisewide project management methodology be established and followed. If there is a separate quality assurance organization, that department performs process compliance audits to ensure that the PM methodology is being followed correctly. If a separate quality department does not exist, the Strategic PMO will undertake this responsibility. Beyond process audits, quality of the deliverables themselves must be monitored and assured. The project manager relies on the quality department or technical departments for this assurance.
- **Change Control**. A rigorous change control process is a mandatory part of any project management process, and is a primary element of Scope Management. Why does this become an enterprise-level concern? Because as projects function at more of an interdepartmental level, the question of the legitimacy of scope changes must be decided from an enterprise level, i.e., taking the entire organization's needs/objectives into account rather than just the sponsoring department. Only a Strategic PMO, operating at the enterprise level, has the

necessary "range of vision" to see the effect of changes relating to one project on the entire portfolio of prioritized projects.

- **Human Resources**, We covered skill inventories and accessibility of salary information, under **System Interfaces** above. Beyond that, project management must be recognized as a career within the organization. There must be a career path for project managers, complete with competitive salaries and opportunities for professional development and promotion. Otherwise, project managers will consider project management as an additional duty to current technical responsibilities. To get promoted, they will be compelled to return to their technical departments. Project managers must have a clear promotion path to senior positions by excelling in the practice of project management. Also within the area of HR management, project managers must understand how to motivate people and teams, not just manage the projects.
- **Communications**. Communication is tough enough when projects are managed individually or within a single business unit. But, when projects are managed as part of an interrelated portfolio, it is necessary that communication take place with *all* the stakeholders in the enterprise simultaneously. Although it's not necessary or desirable for all communication to pass through the PMO, certain key communications should be monitored by the PMO, such as kickoff meeting notes, copies of status reports, lessons learned reports, and other data that will become part of the project management knowledge base. Each individual project plan should include a communication plan, and the project management methodology that governs project management across the enterprise should clearly state the minimum requirements for communicating with other stakeholders and the Strategic Project Office.
- **PM Methodology**. When we left our imaginary case study, Merchant SuperStores had 15 project managers working on 150 projects annually. Assume for the moment that each project manager was governed only by the triple constraints of time, cost, and technical performance, and had no other process

direction. It is reasonable to assume that any given customer of these 150 projects could experience five to ten different project-phase schemes, different review/sign-off requirements, different scheduling mechanisms, different control mechanisms, and a variety of team structures and project management processes. And, as resources move from project to project, they will have to learn the unique processes, tools, and techniques in place on each project. More time will be spent trying to figure out *how* to manage and participate in projects than on the technical and creative work of solving problems and delivering value to customers. This is obviously unacceptable, yet this very situation exists in many real-world enterprises. In order to prevent this situation and bring some order to the chaos, a Project Management Methodology must be created, deployed, and enforced across the enterprise. A good PM Methodology will take into account each of the nine *PMBOK Guide* knowledge areas, and will detail what is to be done and how to do it for each activity in the project life cycle.

- **Procurement**. As indicated, many projects will require the procurement of hardware, software, and other materials and services, and some projects will be performed with outside vendors. It is imperative that project budgets and actual costs be determined and documented correctly, and that this information is concisely reflected in specifications and statements of work in vendor contracts. The vendor selection process is usually governed by methodology within the Procurement Department, not the PMO. The project manager or director of the PMO should ensure that the process selects the best vendor for the project in question. The individual contracts themselves should have clear performance requirements and a method for measuring contractor performance, as well as specific requirements for progress reporting, tied closely to progress payments to the contractor.
- **Training**. Some project managers are smart or experienced enough to "wing it" and still achieve success, but the majority will do a better job if they are properly trained. Knowledge can be transferred through formal training courses, a mentoring

program, project workshops, college courses, or the "school of hard knocks," otherwise known as experience. It is particularly important that the PMO provide training to all project managers, team members, and senior managers on the PM methodology developed for the organization, project leadership, the project team's role in contracting and negotiating, project budgeting and controls, along with the many additional soft skills that support effective project management. This training should be delivered in formal training courses, using extensive hands-on cases and exercises. If a formal project management training program is already underway when the methodology is developed, the courses in the curriculum should be tailored to include the specific activities, templates, forms, and practices called for in the methodology.

- **Cost.** It is particularly important to elevate the coordination of cost management to the strategic level. Most project cost management is concerned with the cost of the resources needed for the project's activities. Because those resources come from multiple departments of the organization, enterprise-level coordination is necessary. However, there are other costs incurred in a project: hardware, software, vendors, and the like. Management techniques, such as return on investment, discounted cash flow, economic value added, and payback analysis, must be considered when managing a portfolio of projects across the company. The PMO Steering Committee, to be discussed later, takes a direct interest in selecting and prioritizing projects based in large part on their return to the organization. Integrated cost estimating, forecasting, and management must be available to the Steering Committee to make these decisions.
- **Risk.** Identification and management of risk is not a one-time event, but must continue throughout each individual project's life cycle as well as across the entire project portfolio. Not only should internal project risks be identified and managed, but also external risks, such as changing market conditions, project funding by external sponsors, political events, environmental changes, and others. These external and cross-project events are clearly strategic, enterprise-level considerations.

Many "Ps" Make a Strong Business Case

I started my work with PMOs years ago with the catchphrase that there were three Ps important to the project office:

- Projects (and Programs)
- Processes
- People

But, as time goes by, I have to keep expanding that list. The scope of the PMO has moved beyond the management of projects and so we add the selection, management, and business alignment of the:

- Portfolio

The PMO must add value, not in terms of project management metrics alone, but in terms of overall business value. Even the PMOs that have excelled at instituting a project management culture and at improving the delivery of projects and programs are in danger of being cancelled if they cannot display the benefits realized by their work, and so we add:

- Performance measurement

Once the PMO becomes engaged in measuring performance against business goals, it is in an excellent position to become the enterprise's center for:

- Pricing

How does an organization determine what to charge for its goods and services? In the past, mistakes have been made in this area because all the costs of creating a new product or service were not considered. As the PMO oversees the estimating of projects, and implements project management cost controls, it gains the ability to accurately price the enterprise's work. At the same time, by tracking costs precisely and allocating the costs of project management back to the business units, the PMO is relieved of the constant need to "justify its existence" in order to gain funding. This brings to mind the final P:

- Positioning

Not long after project managers began pushing for project offices, they realized that merely listing the project management benefits (improved schedule compliance, etc.) was not going to be enough. They would have to learn how to *market* themselves and their services to the organization as a whole, and particularly to senior executives. To judge by the rapid rise of PMOs from inside divisions and business units to enterprise status, we have become remarkably skilled at this. However, even successful Strategic PMOs cannot rest on their laurels. We have seen a number of excellent, even award-winning PMOs suffer cutbacks. As PM Solutions' in-house editor has observed, this is because good project management is a lot like editing: invisible. You can't see the mistakes and problems that aren't there, and so the PMO is always in danger of looking like mere overhead in an organization where everything is running smoothly. The PMO must never cease working to foster internal adoption and acceptance of new projects and programs. They can't stop articulating value of projects and programs to the business. And, as they promote the PMO as a "strategy execution engine" (for all the reasons laid out in Chapter 1), they will be positioning themselves for future success.

Having made the argument for enterprise-level project management in complex organizations, let's move on and look at the optimal organizational structures for growing and nurturing a PMO.

Organizational Structure for Projects

The cross-functional team is a central feature of project management. However, research has shown that simply creating teams within an existing functionally oriented organization will not bring about the advantages they are noted for, such as faster product development and higher success rates with new products.[1] Organizations that organized around functional areas or a weak matrix approach were less successful than projectized teams and teams employing strong matrix structures. These types of organizational structures worked equally well for both complex and simple projects (see Table 2.1 for a comparison of organizational structures).

This isn't that surprising because organizational structure is really all about how decision-making authority and responsibility are allocated. In a matrix, even though cross-functional teams may exist, the functional manager still has the authority over resource assignments and the power to eventually cripple the effectiveness of the team.

Table 2.1 How Projects Fit into Organizations

PROJECT CHARACTERISTICS	ORGANIZATION TYPE				
	FUNCTIONAL	WEAK MATRIX	BALANCED MATRIX	STRONG MATRIX	PROJECTIZED
Project Manager's Authority	Little or None	Limited	Low to Moderate	Moderate to High	High to Almost Total
Percent of Performing Organization's Personnel Assigned Full Time to Project Work	Virtually None	0–25%	15–60%	50–95%	85–100%
Project Manager's Role	Part time	Part time	Full time	Full time	Full time
Common Titles for Project Manager's Role	Project Coordinator/ Project Leader	Project Coordinator/ Project Leader	Project Manager/ Project Officer	Project Manager/ Program Manager	Project Manager/ Program Manager
Project Management Administrative Staff	Part time	Part time	Part time	Full time	Full time

Many companies have already realized the inherent problems of fitting cross-functional teams into existing functional organizations and concluded that it will not work.[2] The old functional culture tends to dominate, making accountability difficult and decision making slow. In forward-thinking companies, one solution has been to empower the program (or projects) by moving to a more projectized structure and by creating groups or centers of expertise (such as the Project Office) to service the programs. This changes the functional managers' role from owner of resources to supplier of people.

Using the project or program as the basic structure of the organization has been a key to success at companies such as Microsoft, Intel, Texas Instruments, AT&T, EDS, Toyota, Ford, and others.[3] Motorola's team-based satellite communications division realized a revenue-to-employee ratio of $2 million to 1 as opposed to a ratio of $150K to 1 for other divisions in the company.[4] These companies often use projects to drive change in the organization and add core capabilities as natural byproducts of the process of delivering products. In a team- and project-based organization, project delivery is enhanced. Table 2.2 shows some of the key attributes of a "projectized" organization. It's no surprise that

Table 2.2 Features of the Projectized Organization

Features of the Projectized Organization
Teams are accountable for performance.
Functional managers are resource suppliers, now resource "owners."
Individual performance reviews are suspended—team performance is key.
Professionals have a say in their assignments.
Teams, not just managers, have access to senior management.
Everyone is trained for team-based work.
Team members stay with their team for the duration of the project.

Source: Adapted from Anne Donnellon, "Cross Functional Teams in Product Development: Accommodating the Structure to the Process," *Journal of Product Innovation Management* (November 1993), 377–392.

most, if not all, of these changes can be initiated as part of the implementation of a PMO, and then effectively managed by that PMO once they are in place. The PMO becomes the facilitating body for cross-functional team management. It provides leadership training to project managers, provides training to the project teams, provides the guidance and direction on how to manage and direct cross-functional product/project teams, and generally helps to create an environment within the organization that is team-friendly and team-optimizing.

Culture change of this magnitude, however, is a tall order. Realistically, most companies begin by shoring up the features of projectization an organization already possesses, and working from there toward the ideal—an enterprise-level project office within an organization where projects and teams, rather than functional departments, are the central organizational units. An exception is in companies that naturally operate as large PMOs—companies where projects are the lifeblood of the business, such as construction firms and professional services firms of many types. A study carried out by Christopher Sauer of Oxford University in the United Kingdom in 2001, noted that such firms have done away with PMOs "… because they *are* PMOs, and all business practices, from strategy management to HR, revolve around projects and teams." He makes a convincing case that any company in any industry can learn much from this paradigm, and that we may all be headed in this direction.[5] His research bolsters the opinion of management guru Tom Peters, who proclaimed that the logic of projectization is "unstoppable" and recommended every company operate as a projectized professional services firm.[6]

Where is your company on this evolutionary scale? Let's examine in detail the features of each stage in the evolution of the PMO.

Types of PMOs

Not all Project Management Offices are created equal, although almost any form of PMO will jumpstart incremental process improvements in organizations that have nothing at all in place. Basically, a PMO is an "office" (either physical or virtual) staffed by project management professionals who serve their organization's project management needs. It also serves as an organizational center for project management excellence. A PMO may exist at any one of three levels in the organization, or PMOs may exist at all three levels concurrently.

Type 1: The Project Control Office This is an entity that typically handles large, complex single projects (such as a Euro conversion project or the creation of a new type of airplane). It is specifically focused on one project, but that one project is so large and so complex that it requires multiple schedules, which may need to integrate into an overall program schedule. It may have multiple project managers who are each independently responsible for an individual project schedule and, as those schedules, their associated resource requirements, and their associated costs are all integrated into an overall program schedule, one program manager or a master project manager is responsible for integrating all of the schedules, the resource requirements, and the costs to ensure that the program as a whole meets its deadlines, milestones, and deliverables.

Type 2: Business Unit PMO At the divisional or business unit level, a PMO may still be required to provide support for individual projects, but its challenge is to integrate a large number of multiple projects of varying sizes, from small, short-term initiatives that require few resources to multimonth or multiyear initiatives requiring dozens of resources, large dollar amounts, and complex integration of technologies. The value of the Type 2 PMO is that it begins to integrate resources at an organizational level, and it's at the organizational level that resource control begins to play a much higher-value role in the payback of a project management system. At the individual project level, applying the discipline of project management creates significant value to the project because it begins to build repeatability—the project schedule, and the project plan, become communication tools among the team members as well as within and among the organizational leadership. At Type 2 and higher, the PMO serves that function, but also begins to provide a much higher level of efficiency in managing resources across projects. Where there are multiple projects vying for a systems designer, for example, the Type 2 PMO has project management systems established to deconflict that competing need for a common resource and identify the relative priorities of projects, such that the higher priority projects receive the resources they need and lower priority projects are either delayed or canceled. A Type 2 PMO allows an organization to determine when resource shortages exist and to have enough information at their fingertips to make decisions on whether to hire or contract additional resources. Because the Type 2 PMO exists within a single department, conflicts that can't be resolved by the PMO can easily be escalated to a department manager, who has ultimate responsibility for performance within his or her department.

Type 3: Strategic PMO Consider an organization with multiple business units, multiple support departments at both the business unit and corporate level, and ongoing projects within each unit. A level 2 project office would have no authority to prioritize projects from the corporate perspective, yet corporate management must select projects that will best support strategic corporate objectives. These objectives could include profitability goals, market penetration strategies, product line expansion, geographic expansion, upgrades to internal information management capability, to name just a few. Only a corporate-level organization can provide the coordination and broad perspective needed to select, prioritize, and monitor projects and programs that contribute to attainment of corporate strategy, and this organization is the Strategic PMO. At the corporate level, the Strategic PMO serves to deconflict the need for competing resources by continuously prioritizing the list of projects across the entire organization. Obviously, this cannot be done by the SPMO in isolation; thus, the need for a steering committee made up of the SPMO director, corporate management, and representatives from each business unit and functional department. The Project Office Steering Committee (the role of which is discussed in more detail in Chapter 7) looks at the contribution of each project to corporate and business unit goals. It also ensures that overlap and integration of projects are considered. For example, one well-known organization found that when it prioritized its list of 103 projects, some projects could not succeed unless other projects failed due to competing technologies. Without an integrated, prioritized list of projects, this would not have been discovered.

The Strategic PMO operates at the appropriate level to facilitate the selection, prioritization, and management of projects that are of corporate interest. It ensures the project management methodology is tailored to the needs of the entire organization, not just one department or business unit. It is important to note in passing, however, that frequently a single methodology is not appropriate for every business unit. For example, one of PM Solutions' clients is in the petroleum exploration and processing business, as well as specialty chemicals, retail operations, asphalt production and paving, and associated businesses. It also has a large, sophisticated information technology department that supports the organization and a large facilities management department. In attempting to develop a methodology that fit the needs of its entire corporate membership, it found that a significant amount of tailoring was required at the lower levels of detail in the methodology to be consistent with industry practices in their various business lines. For example, the facilities department worked with construction and its associated processes. The IT department followed software development and systems integration processes. The oil business likewise had its industry-specific project management terms, forms, templates, and practices. The overall project management process groups—initiating, planning, execution, control, and closing—applied to all projects, but many of the forms and templates supporting these processes were tailored to the business units.

Which type of project office is right for your organization? It depends on the size and complexity of the company, the interdependence of projects among business units and functions, the availability of resources, and the competence of your project managers, among other things. Large, complex projects need a core team of talented individuals to manage them. This can be considered a Type 1 project office. However, if you have several complex, important projects, a

Type 2 PMO is called for to capture and institutionalize the best practices of the best project teams. And, if the organization is large, with scarce resources and many critically important projects, a Type 3 project office is needed to ensure that corporate strategy is realized through the most effective and efficient execution of projects.

Regardless of which type of PMO you envision for your organization, all PMOs perform the same functions to one degree or another. These core functions are reviewed below.

Functions of the Project Management Office

There are six primary components to any PMO, which grow in capability and complexity as the PMO takes on more strategic responsibilities.

Project Support

There's a significant element of project management that requires project planning, project scheduling, cost control, administration, controls, and other detailed technical tasks—what we call the science of project management. However, a much more important segment of the project manager's work deals with the art of project management, areas such as leadership, negotiation, motivation, teambuilding, facilitation, analysis, project chartering, incentive creation, and the like. To provide the appropriate level of technical support for project managers so that they can focus on the things at which they can have the greater impact is an important role for the PMO. Project support is a specialty focus that requires considerable expertise in the specialty areas: scheduling and project controls. These skills may take time to develop, are critical to retain, and adhere to a standard process and approach specific to the project support function.

Documentation The project support group performs the science of project management, as opposed to the art. They are responsible for estimating and budgeting, including cost estimating and capital estimating. They develop plans and schedules. Project support works with project teams to develop a work breakdown structure (WBS), follow that structure with the network diagram, get the network diagram into the project's scheduling software, have the scheduling software run the forward and backward pass to identify the critical path and float, and, with some additional analysis, produce the final schedules. In addition, they also provide status updates, pulling data from time collection, timesheets, and the financial system to update the status against the plan. They perform variance analysis, and all the associated reporting both back to the project team and also up through management and throughout the organization.

There's a significant amount of data entry involved in project support, but it is data that is the lifeblood of project control. In some cases, the status of critical projects must be updated daily, with changes to the schedule and modifications in the "visibility room" (once known as the "war room") so that key stakeholders who need the updated information can see exactly where the project is in real time.

Change Control Project support functions are critical to change control. Each change must be documented on a change request form, including analysis of impact on cost, schedule, and technical baselines, as well as disposition of the change. Minutes from the change control board must be recorded and disseminated. A change log must be maintained showing status of all changes, whether approved or disapproved. All this must be managed as part of a project manager's job. The PMO can play a key role, and significantly eases the paperwork burden on project managers by managing the change control process. The PMO will convene change boards and take care of the resulting documentation. It ensures that approved changes are reflected in specifications and contract documents, and that team members who need to be kept abreast of changes are notified in a timely manner.

Project Repository Project support also entails keeping a project repository, which may be as simple as a book or as complex as a knowledge management system. It's a record of all the documentation, all the plans—an historical record to reference in the event that a project manager or team member leaves the project. This project repository helps indoctrinate new project managers to lessons learned and enables a smooth transition into and out of new and old projects.

Tracking and Reporting Project support also maintains issues tracking. Much like change control, issues and action item tracking can be substantial depending on the number of projects and the number of people in the organization.

Project progress reports roll up into summaries for the appropriate functional areas, are further summarized for appropriate levels of the matrix organization, and so on up to executive management. Thus, the project support organization is responsible for *the executive dashboard.* That is a fancy way to say that they are responsible for executive reporting, but the dashboard typically is focused on keeping a reporting structure that is distilled to one page. Executives should look at one page of information and see that projects on a whole are on schedule or not, and if they are not, where they can go to find out the issues that are creating problems for the projects. Therefore, the executive dashboard might be an electronic or a paper report, but, in all cases, it needs to be a summary type: succinct, precise, and focused specifically on the information that executives need for decision making.

Risk Management Project support also includes risk management. Risks in each project must be identified, analyzed, mitigated, and followed up. Plans for how to handle each high and moderate risk event must be put in place and action taken to ensure responses are executed on a timely basis. In some cases, where the risk may be so significant as to require rebaselining the project plan, the project office works with the project manager to develop alternatives that might affect the schedule, budget, or project scope. Because risk analysis tends to be a specialty area, in terms of both software and techniques, risk analysis experts typically reside in the project support area of the project office. They provide software and technical support to project managers and frequently help to facilitate risk planning and replanning sessions with project teams.

Resource Repository An organization of any size must maintain a resource repository. The resource repository is an inventory of all available resources throughout the organization or throughout the enterprise. As individuals are added to the company, their profile is established in the resource repository and updated as they develop new skills. The project office, especially at Type 3, acts as a resource broker with functional heads to ensure the right resources are working on the right projects at the right time.

Cost Tracking In organizations with mature project management processes, the accounting system may have been altered to provide actual cost data directly to the project manager via the project management software. However, this is rare. Accounting systems are slow to change due to the expense involved as well as the simple fact that they are entrenched bureaucracies themselves. If the organization does not have direct access, through integrated enterprisewide software, to cost information at the project level, a lot of legwork is required to provide project managers with current, accurate, and complete cost information. Without this data, any attempt at assessing current and projected cost variance is an illusion. This is where the project office comes in. Members of the project support team within the project office literally "mine" the information they need from available data sources. They ask questions, visit the procurement department for information on contract costs, dig out actual hours worked from time-keeping systems, talk to resources to validate information—literally go anywhere in the organization to get the needed information. Individual project managers simply do not have the time to do this. As a result, without a project office providing this support, they have no clear idea of cost variance on their projects.

Software Support Finally, project support handles all issues surrounding the project management software; a category of responsibility so large we have named it one of the central components of the PMO.

Software Tools

The PMO centralizes the establishment and maintenance of project-related software tools. With that also comes establishing and maintaining project management software standards, which will lead the organization into establishing common coding structures so that project schedules can be integrated and rolled up for management-level resource, cost, and schedule reporting.

Project support is responsible for acquiring project management software and supporting software, which could include time collection software, time reporting software, configuration control software, documentation software, library software, database software, spreadsheets, and other software applications. The project support group identifies the needs in the software area, facilitates or performs the integration and use of that software, and then maintains and monitors its performance.

This project support group makes the ultimate decision on which project management software the project manager would use as well as the decision on risk management software or integration software so that the tools used across the enterprise are compatible with each other. This is necessary so that resources and costs can be summarized at the enterprise level.

Project support also establishes a project help desk where project managers or other project team members can receive specialized or expert help with the project management or risk management software tools. They may need help with templates, database design, database application, spreadsheet development, spreadsheet use, template development, and so forth.

Processes, Standards, and Methodologies

The PMO is also responsible for developing and maintaining processes and methodologies pertaining to the management of projects. It serves as a central library for these standards, and is the expert on their deployment. The project office also incorporates lessons learned on projects nearing completion into the PM methodology.

If the PMO is sizable enough, there will usually be a separate individual responsible for process development and maintenance. At a minimum, a project management methodology is developed, but other processes for which the project office may be responsible include the systems development life cycle, software development life cycle, process development life cycles, and product development life cycles. This area of the PMO also functions as a keeper of standards, whether these are industry standards, such as the *PMBOK Guide*, internal company standards, or standards brought in by a consulting firm.

As the keeper of these standards, the project office also maintains the templates, forms, and checklists developed to ease the paperwork burden on project managers. In many cases, the templates are an integral part of the standards themselves. The project management methodology may require, for example, a risk plan, and project managers are helped considerably if they can see a sample of a risk plan including instructions for completing one. Templates, such as the project request form, project charter form, communication plan, risk log, issues log, templates for schedules, templates for cost estimates, templates for the resource assignment matrix, and so on, should be standardized across the enterprise. If basic processes are standardized, comparison and prioritization of projects becomes much easier.

Part of maintaining methods and standards is continually looking at industry to determine best practices. For example, one leading knowledge technology company did a fantastic job of establishing best practices for the product development industry. They looked at the different professional organizations and came up with a summary of best practices for new product development: Use a stage gate process, fully integrated with a comprehensive project management process. Benchmarking individual projects against other projects of like kind either within the organization or within the industry also has immediate practical application. For example, in estimating the deployment of a particular software package, rather than just pulling an estimate out of the air, the project manager can look at other organizations that have deployed a similar software package under similar conditions. Benchmarking helps to legitimize project cost, schedule, and resource requirements estimates.

Benchmarking also can take the form of contracting with a consulting firm known for expertise in your area of interest. For example, in developing a project management methodology, it makes sense to contract with a firm that has developed methodologies for other firms, both in your industry sector and outside your sector. In this way, an organization is purchasing a best practices knowledge base as part of the contract price with the consulting firm.

Whether benchmarking is done in-house or acquired from industry and association analysts, such as the Project Management Institute, Gartner, Inc., MetaGroup, National Contract Management Association, or Forrester Research, among others, the acquisition and application of industrywide data to validate whether or not your particular processes fall within parameters commonly accepted in industry is a primary responsibility of the PMO.

Building a world-class set of processes and methodologies also involves taking advantage of the lessons your own project managers learn while engaged in projects. An archive of lessons learned and methods and processes documentation is one of the PMO's key contributions to standardizing methodology across the organization.

This library of information and data is assembled from past projects: what worked, what didn't work, and how it can be reapplied more effectively to other projects within the organization. One strategy of the software development process engineers has been to develop a process database for software development methodologies. They developed partners in the business community, customers to whom they had sold a component or process tool, and as these partners used the tools and built their own templates, those templates were then provided back to the vendor who standardized them and put them in their library of approaches to planning and managing projects. In their process database they now have methods for implementing client/server systems, for prototyping, for iterative development, for Lotus Notes deployments, SAP deployments, and so forth. Keeping a library of methods used on projects that can be readily reapplied is a timesaver and also can serve as part of the quality function.

The methods and standards area of the Project Office also may serve as the quality audit and continuous improvement function for project management because they understand what should be done in terms of methodology and process and can audit against whether or not it is being done and, if it is being done, whether or not it is showing value and productivity.

The Project Office should also establish and manage an intranet or Web presence through which the standards and templates can be exchanged. Plans, estimates, schedules, deliverables, and project status can be integrated and communicated through a Project Office Web site or intranet site.

Finally, the Project Office methods and standards group works with the organization to manage the project portfolio. They help to establish a portfolio prioritization process by providing facilitators or consultants to structure and administer that process with the executives. They document the results of prioritization within the division or within the company and communicate those results to the planning teams and integration teams. And, they prepare the ongoing, regularly scheduled review that validates the portfolio. They ensure the portfolio is revisited, reviewed and revalidated, and reprioritized as necessary, and that those priorities flow over into the project planning and control process. (For more on project portfolio management, see Chapter 9.)

Training

The Project Management Office is the center of focus for project manager and team training. It identifies competencies needed by high-performing project managers as well as for executive awareness and team member participation. The PMO participates, with a specialized project management training vendor typically, in tailoring standardized courses around the culture and methodologies that specifically apply to the organization.

Although the training department will be the coordinator of corporate training, the PMO provides subject matter expertise in project management. The PMO identifies the appropriate training that is required and participates in selection of the trainers. The PMO also identifies the required levels of knowledge and competency and the required segments of training that are necessary in order to achieve maximum performance. Thus, the PMO is the focal point for measuring project manager competency. (For more on project manager competency, see Chapter 7.)

Consulting and Mentoring

When another department in the enterprise—marketing, for example—wants to manage a project themselves, the Project Management Office can provide expert assistance in the form of counseling and coaching for the staff involved in these projects. This component also provides an audit function for existing and ongoing projects to determine how effectively the project management process is being utilized and deployed within the organization.

The entire organization should view the PMO as a source of specialized experts who have focused concentration and ability in project management. The leader of the marketing department's event planning team, who may not be experienced in project management, can go to the PMO for startup assistance and advice. PMO consultants, or coaches or mentors (the language used varies), provide counsel and conceptual understanding of what's necessary. They clearly understand the science of PM, but they also would have a very solid understanding of the art of planning and managing projects, so they are able to give advice on teambuilding, leadership, communications, negotiations with clients or vendors, problem solving, facilitation, etc. Mentors may not be called on until a problem or challenge arises, so they should be very experienced project managers. They must know the nuts and bolts of project management, such as how to draft a project charter, develop and design the project plan, develop a risk plan, and rescue a project in distress. But they also must know the business and the industry. When a project is facing challenging times and needs additional oversight and guidance to achieve project recovery, the mentor can help develop the project workaround plans, the new estimates of cost, resource reallocations, and replanning.

In addition, a project expert frequently works with a project manager and marketing manager to develop proposals for external work. Because proposals are sometimes accepted without negotiation, estimates of time and cost for the scope proposed must be accurate. This necessitates a greater level of planning earlier in the project life cycle than might be customary. If project planning is sound and estimates are accurate, profitability on contracted projects can be predicted with a reasonable degree of certainty.

In sum, the role of project mentors and consultants is to transfer the knowledge they have developed to project managers and project teams to enable them to perform better on current and future projects. Knowledge transfer is the key because mentors are not provided on a continuing basis.

Finally, mentors and consultants from the project office are the logical personnel to do project assessments or audits. They develop the project audit process and checklists, help determine the timing of the assessments, and conduct the audits. Following the audit of a project in trouble, the auditors might themselves be the most logical experts to help the project manager get back on schedule or within budget.

Project Managers

A stable of capable, qualified, professional project managers can manage a full range of projects, from large, complex technology projects to smaller, short-term projects. To remove the functional bias that inevitably creeps into the project manager's psyche, these professionals are sometimes assigned full time to the project office. In the fully deployed Strategic Project Office, project managers actually report to the PMO and are deployed to projects either as full-time managers or on a part-time basis. The PMO maintains a database of project managers, their skill sets, capabilities, specialties, experience, and technical skills. Using this database, the Project Office Steering Committee discusses the needs of new projects and, in consultation with the director of the project office, recommends assignment of specific project managers to specific projects. Project managers between full-time assignments work on special projects, such as developing new processes, methodologies, techniques, templates, and capabilities. A highly competent project manager is too valuable be idle just because he or she is between projects. This infrastructure development aspect of the PMO allows the organization to derive full value from a project manager's expertise and experience as well as avoid the high cost of turnover.

In the organization of project managers, the manager of project managers or director of the PMO would be responsible for evaluating, coaching, and developing project managers and project team performance. He or she also would be responsible for managing the database of projects and project assignments and matching project requirements with project managers.

Integration of the Project Management Office

Since major projects usually cross all organizational divisions, the most effective use of the project management expertise resident in the PMO will be made if the PMO is integrated into activities, such as project prioritization, budgeting, and cost allocation at the corporate level, where the strategic decisions are made as to how to allocate resources.

In other words, as projects ebb and flow, start, ramp up, and finish, the Strategic PMO is the place where project data is collected and, thus, is the place in the organization best suited to track corporate trends. As they collect data on resource skills, on project tracking, and on training needs, they have data to make decisions on hiring and contracting. The Strategic PMO is the umbrella for project needs, much like the finance department acts as the organization's treasurer and accounting office. Even within a company that has a Strategic PMO, there may be individual projects that are managed internally within a single business unit, of course; however, those individual projects will have the benefit of being standardized, supported, and mentored if the need arises by the SPO.

Evolving PMO Functions

In addition to the six areas discussed above, and the common functions displayed in Table 2.3, as PMOs increasingly become partners in strategy management, further areas of responsibility are emerging. For example, leading edge PMOs now:

- Enlarge the breadth of PMO influence to extend from strategy formulation through benefits realization.
- Design governance to focus senior management on strategic issues.
- Integrate benefits realization into the entire life cycle, starting with planning, and report on it regularly.
- Implement portfolio management tools that provide high-level visibility and analysis that inform decision makers and evoke action.
- Broaden PMO staff competencies to include strategic planning and investment analysis.

Table 2.3 Common PMO Functions

Project reporting and tracking	78.3%
PM process/methodology implementation	77.3%
PM software tool support	64.4%
Project support (planning, scheduling, etc.)	62.7%
Project manager and team training	58.3%
Managing change control	58.0%
Capturing lessons learned	54.6%
Assessing/auditing projects	52.9%
Coaching/mentoring project managers	51.2%
Resource allocation process	46.4%
Managing project managers	46.4%
Prioritizing the project portfolio	42.4%

Source: *The State of Project Management 2006* (Glen Mills, PA: PM Solutions' Center for Business Practices, 2006).

Notes

1. Per a best practices study of 189 firms by the Product Development Management Association. Cited in Clark and Fujimoto, "The Power of Product Integrity," *Harvard Business Review* (November–December 1990). See also Ann Donnellon, "Cross Functional Teams in Product Development: Accommodating the Structure to the Process," *Journal of Product Innovation Management* (November 1993), and Robert G. Cooper, *Winning at New Projects: Accelerating the Process from Idea to launch*, 2nd ed. (Reading, MA: Addison Wesley, 1993).
2. Cited in "Case Study: DFM, Simultaneous Engineering, Employee Involvement Aid Deere," *CIMWEEK* (January 28, 1991). See also Cole, Clark, and Nemec, "Reengineering Information Systems at Cincinnati Milacron," *Planning Review* (May 1993), and Christopher Lorenz, "Reaping the Harvest of an Integrated Ream Approach," *Financial Times* (October 7, 1991).
3. Cited in William Bridges, "The End of the Job," *Fortune* (September 19, 1994). See also Tim Davis, "Reengineering in Action," *Planning Review* (July 1993), Katzenbach and Smith's book, *The Wisdom of Teams: Creating the High Performance Organization* (New York: HarperCollins, 1993), Tom Peters, *Liberation Management* (New York: Knopf, 1992), and Gifford and Elizabeth Pinchot, *The End of Bureaucracy and the Rise of the Intelligent Organization* (San Francisco: Berrett-Koehler, 1993).
4. Theodore B. Kinni, "Boundary-Busting Teamwork: Motorola Leaps Organizational Borders to Create the Infrastructure of Iridium," *Industry Week* (March 21, 1994).
5. Christopher Sauer, "Where Project Managers Are Kings," *Project Management Journal* (December 2001): 31–37.
6. Tom Peters, quoted in "Passion Beats Planning …," Jeannette Cabanis-Brewin, *PM Network* (September 1998).

3

The Starting Gate

Assessing Your Current Condition

How many projects are underway right now in your organization? How many people do you have with project management skills and knowledge? How is that competence documented? Is there a formal career path established for them? What are the project management needs of all your business units and departments—not just information technology (IT) and the other project-oriented ones, but all of them—and what specific skills do they most need? How does each and every ongoing project now link back to corporate strategy?

If this barrage of questions elicits some head-scratching, don't feel bad. Joel Koppelman, co-founder of the project management software vendor Primavera, estimates that very few companies can even answer the first question. In fact, documentation of project activity is so poor in most organizations that they don't even realize that most of their value-added is generated by the project engine.

Thus, like any process improvement initiative, the drive toward a project office must begin with assessing one's current condition—establishing a baseline. While no industry standard yet exists for baselining the capabilities of an organization's project management functions, there are numerous models out there designed to measure project management maturity, most of them based on the IT-industry standard Software Engineering Institute (SEI) Capability Maturity Model (CMM) for Software Development. Let's look at one of these models, the one developed and used in the field by my own firm,[1] and discuss how maturity modeling assists the organization in answering the questions: How are we doing and do we need a Strategic PMO?

Maturity and the PMO

Maturity modeling for project management is a huge topic, one which has consumed a great deal of energy by many skilled and thoughtful project management practitioners for the past several years. So much is involved in determining all aspects of a fully mature PMO, and what follows should be regarded as no more than a quick study of the subject.

A Maturing Profession

Project management has been known as the "accidental profession." How many PMs can actually say they planned to be a project manager when they were in grade school, or even college? Not many. Chances are that your first project management job involved rescuing some undertaking that was out of control. It was probably not even called a project, and you may have been referred to as a troubleshooter. If you were successful in pulling that first endeavor back from the brink of disaster, you were rewarded with—you guessed it—another opportunity to excel. Before long, other people began to ask you about your keys to success. This process has been repeated across the nation and around the world. As time went on, the literature began to reflect this troubleshooting effort as project management.

Project management expert Paul Dinsmore describes the discipline[2] as one that "grew up from the grassroots" in most organizations rather than being imposed from above. For this reason, its role, benefits, and productivity have been hard to assess. Project management in many firms has been outside the mainstream of measurement and reward systems. Now that research is showing that it is an important factor in competitive success, companies are scrambling to figure out how to make the best of this productive resource.

Here's a quick overview of the thought process behind maturity modeling.

Defining Maturity and Capability

Thanks to a decade of work by the SEI, sponsored by Carnegie Mellon University, we have a much better understanding today of the areas of expertise necessary in order for an organization to consistently

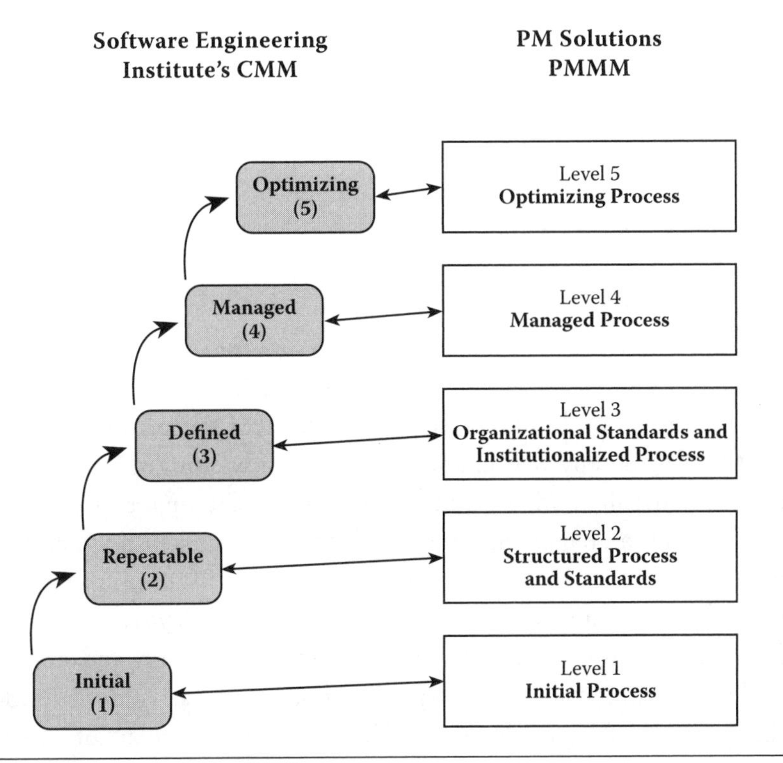

Figure 3.1 Mapping CMM to PMMM.

produce quality software products. The software model, as it described a project-driven business, provided project management with a handy springboard to begin constructing a set of process measures for the discipline. Just as a recap, SEI sets five levels of capability (Figure 3.1), which they define as:

1. **Initial or Chaotic**. The software process is ad hoc; few processes are defined, and success depends on individual heroics.
2. **Repeatable**. Basic project management processes are established to track cost, schedule, and functionality. The necessary process discipline is in place to repeat earlier successes.
3. **Defined**. The software process for both management and engineering activities is documented, standardized, and integrated into a standard software process for the organization. All projects use an approved, tailored version of this standard process for developing and maintaining software.

4. **Managed**. Detailed measures of the software process and product quality are collected. Both the software process and products are quantitatively understood and controlled.
5. **Optimized**. Continuous process improvement is enabled by quantitative feedback from the process and from piloting innovative ideas and technologies.

From experience as a project management consultant, I can verify that the lower the level of maturity, the greater the failure rate on projects. Project management must exist as a repeatable, quantifiable process, found in levels three and above of the maturity model, for there to be any real chance of consistently bringing in projects on time, within budget, and according to customer expectations. The project management methodology must make the measurement of scope, quality, and cost a natural part of running projects. Perhaps most important, the processes must be *institutionalized*; i.e., supported by upper management of the organization and applied uniformly throughout the organization, most easily done under the auspices of a project office. Thus, the measurement of project management maturity "plugs in" to other product development and service delivery process maturity models, whether for software or systems or any other development effort we would care to document and measure.

The challenge that has faced project management practitioners has been to develop a maturity model that is appropriate for all industries, whether it be telecommunications, insurance, banking, finance, or any of the other industries in which project management is applied.

Role of Metrics One of the ways we gauge the maturity of, for example, our children is with numbers. We measure height, weight, SAT scores, and the like. But, as any parent knows, these numbers barely scratch the surface. So, while gathering metrics to describe the project process is useful, qualitative research into how the people side of the organization functions can be just as important to the organizational assessment phase of determining maturity. Luckily, the PM Maturity Model and assessment tool described in this chapter (and in our books on the topic) cover both quantitative and qualitative measures of success.

What are some typical quantitative metrics used to assess project management maturity? For the answer to that question, we go back to the project "triple constraints" of time, cost, and technical/scope. A less-mature organization would consistently miss scheduled milestones and completion dates. This would show up quantitatively as large schedule variances, rescheduled shipments, missed product introductions, and customer complaints. Likewise, cost performance would be poor. This would surface as large cost variances, overruns, requests for additional funds, and shrinking profit margins. And, on the technical/scope side, the two primary measures most frequently associated with lack of process are runaway scope growth and low customer satisfaction. All these are easy to spot during even a cursory audit of project performance within a single project, a division, or the entire organization

Taking metrics to the Project Management Office (PMO) level, the scope broadens. The basic functions of a PMO, such as methodology development, providing expertise to the organization, facilitating project planning sessions, and helping define the corporate software standards, are used to measure effectiveness. A simple yes/no checklist can be used in some cases. For example, does a project management methodology exist? Yes or no? If the answer is yes, you are at least thinking about Level 2 maturity. The next question would be whether the methodology is mandatory or voluntary. If it's mandatory, you may be near Level 3 maturity. If it is voluntary, obviously project performance will be spotty across the organization, indicating a less-mature project office. Another example is the level of training and development of project managers. Has the project office established a training curriculum for project managers? Yes or no? Does it lead to professional certification? And, for a final measure of the effectiveness of the training program, has project schedule, cost, and technical performance improved as a result of the training? Obviously for some of these metrics a sophisticated assessment tool is not needed. However, for a comprehensive assessment of project management maturity, a systematic examination of the practice should be undertaken. More on this later (see Chapter 4).

Be careful that you do not attempt to define organizational maturity and capability in terms of financial and other numerical metrics only, as this may give an unbalanced picture. Numbers can only reflect what has already happened, not why it happened, and not what is possible for the future.

On the qualitative side of the ledger, a company may want to develop metrics to gauge employee satisfaction, customer satisfaction, and stakeholder value. Creating value to stakeholders within and outside of the organization is key to organizational success. Financial measures alone do not present a clear picture of value. They are too unreliable as either a clear gauge of success or a clear picture of the value. Companies that stress shareholder, customers, and employees outperform firms that do not.[3]

Metrics development and tracking is a big subject; we'll come back to it in subsequent chapters when we discuss goal-setting (Chapter 4) and performance measurement (Chapter 10). For now, the important thing to focus your attention on is the fact that doing maturity assessment is a combination of quantitative and qualitative research and, as such, it requires a fairly labor-intensive period of questioning, interviewing, and evaluating people's responses. If the organization were mature enough to have readily available project metrics adequate to describe process maturity, well, you probably would not be reading this book.

One Example of a Maturity Model The Amercan Heritage Dictionary defines *mature* as "having reached full natural growth or development" or "to bring to full development." To define and measure maturity in project management, we must specify levels of growth and development in the requisite skills. Thus, the identification of project manager skills and competencies must be an integral part of any effort to measure the maturity of the processes within which he or she functions. Obviously, this is a complex undertaking.

Our Project Management Maturity Model uses as a starting point and underlying structure the nine knowledge areas of the Project Management Institute's *Guide to the Project Management Body of Knowledge, Fourth Edition* (2008). Why? Because the *PMBOK®* guide contains two frames of reference for addressing the whole project management body of knowledge:

1. **PM Processes**. This approach looks at the management of projects as a set of five tightly integrated, repeatable processes (see Table 3.1 for a full description of these). These processes can be decomposed into a set of activities and tasks necessary to successfully manage a project. This hierarchy of processes

Table 3.1 Project Management Process Groups

1.	**Initiating**: Tasks and activities that conceptualize and/or authorize the project or phase.
2.	**Planning**: Tasks and activities that define and refine objectives and select the best of the alternative courses of action to attain the objectives that the project was undertaken to address.
3.	**Executing**: Tasks and activities that coordinate people and other resources to carry out the plan.
4.	**Monitoring and Controlling**: Tasks and activities that ensure that project objectives are met by monitoring and measuring progress regularly to identify variances from the plan so that corrective action can be taken if necessary.
5.	**Closing**: Tasks and activities that formalize the acceptance of the project or phase and bring it to an orderly end.

Source: Adapted from *A Guide to the Project Management Body of Knowledge*, 4th ed. (Newtown Square, PA: Project Management Institute, 2008).

and activities forms the basis for a project management methodology.

2. **PM Knowledge Areas.** This approach looks at the management of projects as a set of eight interwoven sets of skills/expertise, with a ninth set (Integration Management) that binds them all together. A project manager needs to wear each of these nine "hats" at some time throughout the project and possess knowledge/expertise in all areas, almost as if she or he is nine different, virtual people.

It is this second knowledge-area approach, focusing on knowledge, skills, and expertise, in which the real engine of increasing process maturity lies. In other words, as the project managers within an organization increase their expertise, knowledge, and skills in each of these areas, the organization becomes more *mature* in its practice of project management (see Table 3.2 for a breakdown of these knowledge areas).

Our Project Management Maturity Model (PMMM) decomposes the knowledge areas into their major areas of focus (which we will call "components") and then defines the levels of project management maturity within each component. Because the industry standard Software Engineering Institute's Capability Maturity Model (SEI CMM) for software development is generally used as a guide in the creation of a maturity model, PMMM mirrors the SEI model's five-level structure.

Table 3.2 Nine Organizational Knowledge Areas

- **Project Integration Management**: How well are the various project processes coordinated? Is there a smooth process for making trade-offs among competing objectives and alternatives? Is the work of the project integrated with the ongoing operations of the performing organization?
- **Scope Management**: Are processes in place to ensure that the project includes all the work required to complete the project successfully. Scope management is primarily concerned with defining what is and isn't included in the project work.
- **Procurement/Vendor Management**: Are processes in place for planning the solicitation and procurement of goods and services? For the proper management of the associated contracts?
- **Time Management**: Are processes in place to make sure the project is completed on time? This includes the estimation and scheduling of project activities.
- **Cost Management**: Are processes in place to ensure that the project is completed within the allowed budget? Processes for resource planning, cost estimating, and cost control fall within this knowledge area.
- **Risk Management**: Are processes in place for the systematic identification, analysis, and mitigation or other response to project risks? Both quantitative and qualitative risk analysis are included.
- **Quality Management**: Are processes in place to ensure that the project meets the requirements or satisfies the needs for which it was undertaken? This includes quality policy, assurance, and control.
- **Project Human Resource Management**: Are processes in place to make the most effective use of all the people involved in the project, not only team members, but sponsors, customers, and other stakeholders? This includes identifying, documenting, and assigning roles and responsibilities as well as developing individual and team competencies to enhance performance.
- **Communications Management**: Are processes in place for timely and appropriate management of project information. Determining communication needs, distributing information, and establishing reporting mechanisms fall within this knowledge area.

Source: Adapted from *A Guide to the Project Management Body of Knowledge*, 4th ed. (Newtown Square, PA: Project Management Institute, 2008).

SEI's CMM states: "Maturity Levels 2 through 5 can be characterized through the activities performed by the organization to establish or improve the software process, by activities performed on each project, and by the resulting process capability across projects."[4]

This project-oriented language leads one to infer that direct connections between the CMM and project management maturity can be mapped. SEI goes on to say: "Software process improvement occurs within the context of the organization's strategic plans and business objectives, its organizational structure, the technologies in use, its social culture, and its management system."[5]

In our own work with maturity modeling, we have mapped the levels of the CMM to a model for project management maturity, which we call the PMMM, as shown in Figure 3.1.

Cut to the Chase In practical terms, what all this means is that when an organization wants to improve project management processes, it can start by baselining what exists. It then compares that baseline to an existing maturity model to determine where the organization stands in terms of generally accepted process standards in the project management discipline.

In order to create that baseline study, you break your organization's processes down into knowledge areas, those knowledge areas into components, and each component into a list of tasks, activities, and characteristics about which you can ask a series of simple, practical questions. (See Table 3.3 for a sample of these questions, broken out for two knowledge areas: scope and time management.). The PMMM incorporates thousands of hours of research, years of PMO deployment experience, and lessons learned to give its users the advantage of having been pretested in the field.

Now, let's walk through, in simplified form, the steps of baselining your organization's process maturity.

Preassessment Evaluation

Before making a commitment to engage in a formal project management maturity assessment, your organization should endeavor to gauge its readiness for the process. This is important because it helps establish a baseline for both the organization and any third party from whom you might request assistance in the assessment process. It also sets the stage for determining what you want to receive from an assessment and how you will plan to use the information to benefit the organization and your project management efforts.

For some, this might be an extensive and lengthy process, for others, only an update of what has been documented and tracked. Here are some questions to consider:

- As an organization, what knowledge about project management do we possess?
- How was this knowledge acquired?

Table 3.3 Examples of Assessment Considerations

SCOPE MANAGEMENT

The overall purpose of scope management is to ensure that the project includes all the work required, and only the work required, to complete the project successfully.

REQUIREMENTS DEFINITION/COLLECTION (BUSINESS)

1. The organization uses a standard documented process of gathering and documenting business requirements, which includes obtaining user sign-off of those requirements.
2. The business requirements serve as the basis of project estimation activities (schedule, budget, and other resources).

REQUIREMENTS DEFINITION/COLLECTION (TECHNICAL)

3. The organization uses a standard documented process to translate business requirements into technical requirements.

DELIVERABLES IDENTIFICATION

4. The project has developed a product breakdown structure (PBS), which is integrated into the project's work breakdown structure (WBS).
5. All project deliverables are identified in coordination with client, and quantified in terms that are measurable.
6. All project management products, such as status reports and quality control reports, are included in the PBS.

SCOPE DEFINITION

7. The organization has a standard documented process, including templates, for developing a project charter. The process includes all stakeholders and is used for all projects.
8. Project scope is monitored as part of the project management activity, with deviations being anticipated, documented, and addressed through the change control process/WBS.
9. The project has a WBS with interim milestones and schedules identified that is of sufficient detail to support project planning and control.

SCOPE CHANGE CONTROL

10. The organization uses a defined documented process, which includes all stakeholders and the project plan, to manage scope change. The process defines the forms and approvals that must be obtained prior to the changing of a project's scope.

TIME MANAGEMENT

The overall purpose of time management is to develop the project schedule, manage to that schedule, and ensure the project completes within the approved timeframe.

ACTIVITY DEFINITION

1. All work on the project (business, technical, and management) is included in the project work breakdown structure.
2. Schedule constraints driven by customer, technology, supplier, resource availability, or management requirements are identified and clearly documented.

Table 3.3 Examples of Assessment Considerations (Continued)

ACTIVITY SEQUENCING

3. Dependencies between activities and between products are clearly identified.
4. Organizational standards and templates are used to identify activities, products, and their dependencies.

SCHEDULE DEVELOPMENT

5. The project schedule, including all identified constraints, is developed based upon the initial work breakdown structure.
6. The project schedule is resource loaded and resource leveled. Conflicts are identified, addressed, and resolved.

SCHEDULE CONTROL

7. A schedule baseline is used to measure and report variances between planned and actual progress.
8. The project manager maintains the project schedule on a periodic basis, capturing actual hours and progress metrics for each activity.

SCHEDULE INTEGRATION

9. All project components and subcomponents (e.g., software development, hardware procurement, and subcontractor activities) are integrated into the project schedule.
10. The project schedule is integrated into a higher-level schedule (program or organizational) so that the impact of schedule change can be fully assessed.

- What individuals or group of individuals possess this knowledge?
- What is our "basic" approach to organizing, planning, and managing projects?
- How successful do we think we have been?
- What factors or information form the basis of the measure of success?
- What formal internal assessments have been conducted within the organization?
- Who conducted the assessment?
- What was the focus and content of the assessment?
- What were the conclusions and recommendations?
- Was any action taken based upon the recommendations? (Describe.)
- What support for project management exists within the organizations (both business and IT)? Determine and document the level of support among: senior executives, midlevel management, project managers, and technical staff.

- How does the executive level view the success of current project management practices? What do they say they want changed? In what timeframe?
- Do you believe you will receive support (in terms of people and funding) for entering into a formal assessment process and implementing changes as a result of the assessment and recommendations?
- What risks are associated with the assessment process and implementation of changes?
- What benefits are anticipated from the assessment process?
- Do you want to share any of the internal assessment information with a third party? What information? What are the potential risks of sharing that information, and the risks of *not* sharing it? Are there any contractual considerations? For example, your firm may have proprietary rights in data that must be protected. Suitable contract provisions must be written protecting these rights if you plan to involve consultants or other contractors.

Even though you might consider internal assessment information proprietary in nature, it would be worthwhile to evaluate the risks and benefits of sharing or not sharing internal assessment information before entering into any discussions with a third party about performing an assessment. The advance preparation will help you hit the ground running when and if you do decide to bring in a consultant. Companies sometimes hamstring themselves; in the effort to avoid disclosing too much to an outside party, they disclose too little and wind up wasting time and money working with consultants who don't have the information they need to do a good job. Eventually, you will have to treat consultants doing the assessment as members of the organization and provide full disclosure. Otherwise, the assessment will not be accurate and you will have wasted time and money in the process.

While the preassessment can easily be done in-house, doing a serious assessment of the maturity of your project management processes really requires the skills of an experienced assessor. The assessor should be a certified Project Management Professional, and must understand project initiation, planning, execution, control, and the integration of all knowledge areas in the practice of project management. Further, he or she must be able to accurately assess your level of development and maturity in all these processes and areas. The evaluator must be articulate in writing a development plan for the organization and project office, clearly stating explicit steps that need to be taken to achieve higher levels of maturity. And, he or she must have a grasp of the business and cultural issues within the organization.

Baseline Maturity Assessment

Any assessment of a company or a department's processes—not just project management processes, but any processes—must begin with a firm understanding of the organization's business goals. Without this kind of high-level roadmap, any changes to process or procedure are done in the dark. Unfortunately, as the Balanced Scorecard Collaborative has found in the course of its research into strategic planning, the majority of companies do very poorly at aligning business activities to corporate strategy. In fact, according to their research, 80 to 90 percent of strategies set by top management never come to fruition, in part because they are not communicated to the employees who carry it out and because the company's projects, measurements, and reward system are not aligned with the strategies.[6] If you are considering implementing a PMO, or any major organizational change, you would do well to first inspect the alignment of corporate strategy to the daily activities of employees. Do they have "a firm understanding of the organization's business goals?"

Once it is clear what business objectives the organization is striving toward, the assessment can proceed. We've found that a two-step assessment process incorporating questionnaires and interviews, a process we call a PM HealthCheck, serves very well to uncover an organization's strengths and weaknesses. This process is briefly described below.

PM HealthCheck: Understanding an Organization's Project Management Maturity

The assessment begins with an understanding of an organization's business goals and organizational strengths and identifying areas for improvement. The result of the PM assessment is analysis, documentation, and a PM improvement plan.

The PM HealthCheck incorporates feedback from all stakeholders in the project management process. It is a two-step process of questionnaires and interviews designed to maximize the amount of information and minimize the time commitment of people within the organization.

The analysis and findings are mapped against PM Solutions' PMMM. This matrix illustrates current positioning and the steps necessary to reach project management maturity. It also is used for measuring the progress of the organization's project management initiative.

Deliverables Deliverables from a PM assessment include:

- PM assessment report
- PM improvement plan

The assessment report incorporates goals and information gathered from the PM HealthCheck. The information is mapped against the PMMM continuum, which helps identify gaps and a path for improvement.

The PM improvement plan is a series of action steps and activities needed to close the gap and achieve the organization's desired level of project management maturity. When building the plan, we keep the following principles in mind:

- Build on strengths
- Augment with best practices
- Teach and apply new skills
- Provide appropriate support structure
- Keep it simple
- Integrate back into the organization
- Add value quickly

Participants The PM assessment is led by a senior member of the analysis team. Input from all stakeholders is essential to an accurate assessment. Questionnaires are given out ahead of time and follow-up interviews are scheduled in groups to maximize information and minimize the time commitment of client employees. The PM assessment requires input from:

- Executive management
- Customers/users

- Key middle managers
- Project managers
- Resource/functional managers
- PMO directors
- Team leaders
- Project team members
- Project support staff
- Technical and other support staff

The effectiveness of the PM assessment hinges on getting candid input from representatives of all areas that are involved in the project. PM Solutions will work with the client in initial meetings to identify the appropriate individuals and schedule their time efficiently.

Research The PM HealthCheck questionnaire has nine sections, based on the Project Management Institute's *PMBOK*™ *Guide* knowledge areas.

Findings and the Path Forward Organizational goals and current capabilities are mapped against the PM Maturity Model. The assessment report presents the findings of the PM HealthCheck and the gap between current capabilities and sought-after goals. The gap provides the starting point for building the PM improvement plan.

The assessment and improvement plan establishes the direction for implementing enterprise project management in the organization.

Artifacts to Be Reviewed as Part of an Assessment These items are selected from a set of projects that represent the full range and complexity of organizational initiatives throughout the enterprise:

- **Project Charter**. A document issued by senior management that authorizes the project manager to use corporate resources to fulfill the purpose of the project
- **Statement of Work**. A document describing the expected outcome of the project, including the deliverables to be produced.
- **Success Criteria**. A document describing the factors by which project success will be measured. These could include schedule and budget performance, rate on investments (ROIs) to be

achieved when developed product is deployed, and a measure of user satisfaction.

- **Project Organization Chart**. A pictorial representation of the project team structure. It should include position titles and names of individuals assigned to those positions. It also should identify all user and support personnel. Other related documents are the Responsibility Matrix, a matrix indicating who on the project team is responsible for what, and job descriptions for each position on the project team.
- **Corporate Organization Chart**. A pictorial representation of the corporate structure. It should include position titles and names of individuals assigned to those positions.
- **Work Breakdown Structure**. A hierarchical representation of all work to be performed as part of the project. The document should include the WBS dictionary, with definition of each work element as well.
- **Product Breakdown Structure**. A hierarchical representation of the products to be produced by the project.
- **Estimating Standards**. The estimating guidelines and actual data used by the project management team to estimate the level of effort needed to complete the project. The standards should identify the type of estimation performed (i.e., lines of code, function point analysis, size of documentation) and any tools used in the estimation process.
- **Project Schedule, including baselines**. The documentation that informs all concerned parties when various project activities will begin and end. The schedule should identify who is doing what work and the dependencies within the activities.
- **Weekly Time/Activity Status Reports**. The data used by the project management team to status the project.
- **Last Six Internal Project Status Reports (including Project Variance Reports)**. Any reports produced for distribution internally within the project team that communicate project status.
- **Last Six External Project Status Reports**. Any reports produced for distribution external to the project team that communicate project status.

- **Project Financial Reports, including budget baselines**. The data used by the project management team to status the project's finances.
- **Project Plan, including Cost and Schedule Management Plans**. The documents that identify how the project will be managed and what the final product developed will be. These documents are usually developed during the project planning activities. They should include a detailed description of the deliverables sign-off process. They also should include a discussion of how the project will be closed down upon completion of all deliverables or upon cancellation. The documents are management documents, not technical documents.
- **Product/Software Development Life Cycle Documentation**. Any documentation supporting the development life cycle for the project. This might include items, such as a design document template, guidelines for user documentation, and release notes instructions.
- **Project Communications Plan**. The document that describes how communications (that includes project status reports) will occur.
- **Risk Identification and Mitigation Plan**. The document that describes the project risks, their probabilities, and the strategies that will be used to mitigate these risks. It also should describe how risks are identified and quantified on the project.
- **Requirements Management Plan**. The document that describes how requirements will be defined and managed throughout the project's life cycle.
- **Quality Assurance Plan**. The document that describes how the corporation will evaluate the project's performance to ensure that the project will meet corporate quality standards. The Quality Assurance manager should be identified.
- **Quality Management Plan (including testing documents)**. The documents that describe how the project will meet corporate quality standards. The document should include a discussion of all quality control activities including peer reviews and audits. It should identify those resources specifically charged with quality control activities.

- **Quality Control Plan**. The documents that describe how the project will implement quality control activities including peer reviews and audits. It should identify those resources specifically charged with managing and conducting these activities. As an example, in software development, it should include standards to be used in the activities, including forms, coding standards, definitions of defects and severity codes, plus any reporting standards applicable.
- **Change Control Plan**. The document that describes how changes to the project's scope will be managed. The make-up of the Change Control Board, including how often it convenes, should be discussed in the plan.
- **Change Requests and Log**. The documents used in the project's change control process.
- **Peer Review Reports**. Minutes and notes from peer review sessions.
- **Action Items Tracking Documents**. The documents used to track action items for the project.
- **PMO Charter**. The document that describes the roles and contributions of the PMO.
- **Subcontractor Contracts**. The legally binding document describing the statement of work, including terms and conditions, applicable to subcontractors used on the project.
- **Project Team Training Plan**. The document describing any training to be provided by the project to team members.
- **Lessons Learned Reports**. Any postmortems that have been performed on the project to date.
- **Miscellaneous Project Documentation**. This umbrella category includes the following artifacts: project status meeting agenda, project meeting minutes, kick-off meeting agenda, and minutes.
- **Time Reporting System Standard Operating Procedure (SOP) and Timecard Data**. The documentation that describes how time charged to a job is recorded and used to report project status. Timecard data should be reviewed to determine how actuals have been used to modify estimating standards.
- **Visibility (War) Room and Procedures**. Any documentation supporting the project's "visibility room," if one exists.

Interview Checklists Once the artifacts are gathered, interviews begin, using comprehensive checklists for each knowledge area. An example of two typical checklists can be found in Table 3.3. Upon completion of a review of the artifacts and compilation of interview data, a reasonably accurate picture of the organization's current condition emerges. But the assessment, in and of itself, is just the beginning. The assessment report points you toward an understanding of your maturity level. Fully assessing maturity is a far more complex proposition than most companies who are just looking to launch a project management initiative fully realize. What the assessment does, in a fairly quick manner, is indicate where the immediate problems are and generate a "hit list" of issues that can be addressed to immediately improve the project management processes.

Deliverables and Results The deliverables of an assessment effort, such as HealthCheck, include the assessment report, detailing the quantitative and qualitative results of the survey, and the PM improvement plan. These documents include a summary of the conditions, strengths, and weaknesses within the organization and recommendations for improvement, along with a plan for how to proceed with these improvements. An example of a comprehensive PM assessment report and improvement plan can be found in Appendix C.

It is up to senior management to choose which level of maturity is desired, and how much of the improvement plan to implement. Naturally, the more comprehensive the effort, the greater the cost and time for implementation.

Sometimes, after completing an assessment and reading the improvement plan, an organization is unwilling to devote the time and resources necessary to reach Project Management Maturity Level 5. To progress from Level 1 or 2 to Level 5 will divert scarce resources and management attention away from the primary business objectives. For this reason, implementation plans must be tailored to fit the amount of change the organization is ready to accept, which is usually something less than what it would to take to leapfrog to the highest maturity level. And, in fact, it may not make sense to attempt such an effort in the short term. It's important to keep your eye on the prize, and the prize is better run projects and more repeatable processes, not a plaque with a certain maturity model level engraved on it. The modeling process is merely a tool that allows us to work toward the end of improving processes and thereby improving productivity and, ultimately, profitability.

Identification of Issues and Risks

Now that you have done the assessment and better understand the maturity model itself, you are prepared to look at the key issues and risks that the organization is facing. What issues will have to be dealt with in order to move forward with a project management improvement plan? What risk events may occur that could derail the effort? While many issues and risks will have been identified in the PM improvement plan, the task now is deciding how to deal with these issues in such a way that will ensure a smooth transition to full-scale implementation of project management. Let's look at a case study—The Money Super Market, Ltd. Case—to reinforce some of these issues. Read the case in Table 3.4, then refer to Table 3.5 for a list of implementation issues from a meeting convened for that purpose.

We have found in working with a number of customers that the same issues found in our case study arise in many organizations. The important thing at this point is to invite participation of all stakeholders in the change process, and to conduct the meeting in such a way as to ensure the free flow of ideas. You might consider using your assessment consultant for this purpose to remove any bias that might creep in from the stakeholders. (See Table 3.6 for a few helpful tips on facilitating issues identification meetings.)

Keep in mind that the issues list doesn't die once implementation of the recommendations to improve project management practices begins. Once your implementation plan is developed, you will want to backtrack and do a "sanity check" against the issues list to ensure that the plan has adequately covered all the issues identified in this brainstorming session. If the issues have not been adequately addressed in the implementation plan, the choice may be made to replan or to table the issue for the time being. The issues list is both a planning tool and a quality check.

Table 3.4 Case Study: The Money Super Market, Ltd.

BACKGROUND

The Money Super Market, Ltd. (MSM) is a large financial institution that was founded in the early 1950s. MSM revenues grew conservatively for 40 years under the direction of its founder, Ross T. Nichols. Since the founder's death in 1989, and under new, aggressive management, MSM has undergone major changes in business direction as new opportunities developed as a result of deregulation. However, the majority of MSM's growth came mainly from acquisitions in the 1990s.

Although MSM's financial performance in the 1990s was astounding for investors, it took its toll on the infrastructure as costs rose at a higher rate than revenues. The many acquisitions resulted in duplicative functions and systems. Systems maintenance and enhancement costs consumed 80 percent of the IT $10 million annual budget. In the last few years, investors have grown leery of MSM's ability to sustain its financial performance as aggressive competition erodes its market share.

Today, only 20 percent of the IT budget can be devoted to supporting new business opportunities, resulting in lost opportunities and revenue. The business unit managers are very upset with the IT organization. They see the IT organization as providing little or no value to MSM's business. The pressure was intense on the CIO, Justin Time, to provide more IT support for new business initiatives. Three months ago, he resigned in disgrace.

CURRENT SITUATION

A new CIO, I. M. Miracle, has been on board for one month. Miracle comes from a small but aggressive company, Dollar Deli, Inc. (DDI), which uses project management as a competitive edge. Before assuming the position of CIO at DDI, he led major IT projects to successful completion. He gained his Project Management Professional (PMP) certification in 2002. Miracle is a member of the Project Management Institute and is a strong believer in project management. During this month, he has discovered the following about the MSM IT organization:

- The MSM IT organization has a very poor reputation within the business units.
- There is no communication among the business units regarding project needs and priorities.
- The business unit managers have been politicking to get the COO to outsource the entire IT operation.
- The IT budget will not be increased this year.
- The IT organization has never completed a project within the triple constraints (budget, time, and specifications).
- MSM needs five new IT projects completed in 2000 to remain competitive.
- The IT organization does not have a project management methodology.
- Only a few IT managers are trained in project management.
- The morale in IT is low because of all the derogatory comments from the business units.
- Most MSM project managers came from other companies and use project management tools and techniques that worked for them in the past.
- There are no project standards in place.
- There is no standard estimating process.
- Project risks are never addressed.
- At least three different project management software products are partially used.
- The MSM IT organization is larger than most organizations in companies of comparable size.
- No one in the IT organization knows the total number of projects.
- The IT organization does not track hours or report performance to budget.

Table 3.5 Sample Issues Identification Matrix

MANAGEMENT	ORGANIZATION	PROCESS/ TOOLS
Buy-in from IT department	Diverse Culture	No status reporting
Improve scores in PM HealthCheck assessment	Mistrust between departments	No risk analysis
Not budgeted	Infrastructure too large	No cost tracking
Short time frame to show success	Poor communication	No standard methodology
Low morale	Overall change issues	Duplicate tools
No recognition of a problem	No accountability	No standard for estimating
Management turnover	Culture not conducive	No budget tracking
New CIO (too much, too soon)	Bad image (IT)	No standard PM software
No true project portfolio	No business unit support	No quality control
Inappropriate use of resources	Duplicative systems and functions	Improve procurement process
Unclear expectations	No true project manager	Cannot incur additional expenses
Too many acquisitions	Resistance to change	Do not address risks
No clear roles	No clear roles	No project standards
Deregulation		
Losing market share		

Gap Analysis

A gap analysis is simply a tool to identify the gaps between the desired level of maturity and current capability—basically a statement of where the organization is versus where the organization desires to be The gap analysis helps the organization define a path forward. For each organizational entity shown in the gap analysis chart, present and desired future functions are identified. Key stakeholders in the organization, including the identified project office director or the owner of the project office process, key sponsors, as well as other stakeholders, such as lead project managers who are engaged with active, successful projects, should be included in the group that does the gap analysis. Once each gap is identified, a plan of action is formulated that will address the gap. At the completion of the gap analysis, the overall scope of the improvement effort will be known. Taken with the improvement plan, the implementation team can now define the scope, plan the steps necessary to accomplish the scope and objectives, and begin to execute the plan. If that sounds like project management, it is. Discussed in the next chapter, implementation of a project management maturity improvement plan,

Table 3.6 Facilitation for Issues Identification

- Who should attend: Key participants should include the identified PMO director, the PMO sponsor, key members of the PMO steering committee, and key members of the PMO staff (if they have been identified).
- Provide a "seed list" of common organizational issues related to project management improvement (see Table 3.4 Case Study). The facilitator may want to analyze the data generated by the preassessment and the maturity assessment in order to develop a seed list tailored to the organization.
- As a group, identify which of the issues on the seed list is a reality for your organization. Cross out those issues that don't apply to the organization. (Note: Consensus should be reached on this decision. Sometimes a lonely voice can, in fact, be on target in identifying a problem that others have ignored.)
- Have participants rate the remaining issues on a scale of 1 to 5 from least important to most important.
- Ask participants if there are other areas not on the seed list. Rate these areas from 1 to 5.
- Break the listed issues down into the following categories: management issues, organizational structure or culture issues, process issues, tools issues, training issues, performance issues, and issues related to external support (see Table 3.3).
- Break the meeting into workgroups around the issues and have them brainstorm ways to move forward. A decision not to deal with an issue is valid in some cases. The group as a whole may make a decision, for example, only to deal with issues rated 4 or 5.
- Each workgroup brings its results back to the larger group for further discussion and development of an action plan (task list). Further meetings may be scheduled at this time for development of goals and objectives on those issues that are singled out for action.
- Deliverables from the meeting should include both the task lists and a situation analysis based on the problems and solutions discussed by the workgroups.

Source: Adapted from B. Terence Goodwin, *Write on the Wall* (Alexandria, VA: American Society for Training and Development, 1994).

an element of which is the PMO, can be undertaken as a project, using basic project management practices, tools, and techniques.

A completed gap analysis chart for the case study organization is shown in Table 3.7. It illustrates an idea of the types of input one will receive from this exercise.

In a way, the gap analysis requires the organization to develop a kind of double vision: keeping in constant view the longer-term state, the higher level of maturity that is the ultimate goal, while at the same time moving the organization quickly in the short range to an added-value position by improving each of these areas of consideration.

You will be amazed at the energy that is created when the issues, short- and long-term vision, and benefits are on paper and available for all to see. That's why this process needs to be fully documented as part of PMO deployment planning. When the implementation team

Table 3.7 Sample Gap Analysis for Case Study Organization

	PRESENT	FUTURE (3–5 YEARS)
Methods	Nonexistent	Standard PM methodology and process—standards, templates, and best practices
	No standards	Benchmarking
	No risk analysis	Project history
	Ad hoc	Project life cycle methodology
	Chaos (lack of understanding of current situation)	Web site with templates, guidelines, and examples in use
	Multiple tools and techniques	Risk analysis
	No PM methodology	Method for project portfolio management (prioritizing)
	No project standards	Develop project resource pool
	No risk management process	Estimating
	No portfolio management	Lessons learned repository
	No project selection process	
Training	Only a few PMs trained	Internal certification program
	None internal	A curriculum for PM in place
	Some from new hires (CIO)	Career path in place for PMs, including PMP
	Management development program	PMs trained
	No formal PM training	Tracking system to identify who needs training and who has had it
	No PM career path	PMs development program (curriculum)
	Training and development needed quickly	Tools, people skills (leadership/communication)
		Subjects: methodology, software tools, PMI certification, team building, facilitation
		PM handbook (manual)
Consulting	Nonexistent	Startup and planning assistance
	Flat budget, cannot afford	Project audits/reviews
	No mentoring	Rein in runaways
	No PM services provided	Formal mentoring program
		Provide services for other business units
		Establish partnership with externals
		Develop internal competency for consulting
		Evaluations
		Process support
		Assessments
		Manage projects

Table 3.7 Sample Gap Analysis for Case Study Organization (Continued)

	PRESENT	FUTURE (3–5 YEARS)
		Knowledge transfer
		Audit (QA)
Support	Some estimating	Enterprise scheduling/planning tool and procedures implemented
	Project risks not addressed	Status reporting
	No standardized tools; use at least three software products	Resource and skill repository
	No time, performance, cost, or issue tracking	Time and issue tracking
	No project list or database	Current inventory
	No formal support	Documentation management
	Heavy infrastructure	Resource database in place
	Duplicate systems	Systems: cost accounting, estimates, and scheduling
	No common application of support services	Project database
		Project portfolio
		Estimates
		Libraries (capitalize on experience)
		Formalize management support
		Standardized tools and systems
		Templates
		Resource repository
		Project help desk
Managers	Not identified	More than five years experience
	No PMO	PM pool (bench)
	Some PM skills in new hires	PMO in place
	No bench	Trained PMs
	No evaluations	Manager of PMs
	Assignments made ad hoc	PM report to manager quarterly
	No leadership	Organize current PM in teams
	No formal PM job description	Establish future practice
	Poor communication	Create job description
	No bench	Establish communications
	No database	Sponsorship
	No evaluations	Incentives, rewards, compensation established
	Resistance to change	Role definition
		Resource forecasting
		Career path development
		Formal measurement, evaluations, and accountability

reviews the reports and lists downstream, they will be able to validate that their efforts continue to be directed toward the objectives set early in the assessment and planning phases.

Conclusion

A note of caution: Racing up the "stairs" of a model can be counterproductive. Like any other organizational development tool, a maturity model is a tool, a means to an end. As former ABT vice president Edward Farrelly once warned, "a company must use maturity modeling as a way to meet business goals, not as a goal in itself."[7] Bob Lewis, of Perot Systems Corp., wrote[8] that when it comes to an art/science-like product development, we must be cautious not to standardize the life—the innovative, creative spirit—out of the organization through process measurement that becomes a process straightjacket. Like any living thing, the organization needs a maturation process that develops what's best in it without stunting its growth.

An Iterative Process

Because simply by assessing organizational competence in these areas, you will direct attention to them, and create subtle energy for change (the old Hawthorne effect[9] in action), assessing your organization's situation can't be done just once. Once a company enters the self-assessment mode, be prepared for all kinds of organizational changes, including the change to a "learning organization," one that constantly reevaluates its own progress, learns from history, and refines for future growth and opportunity. I like to think of it as a similar pattern to the iterative Software Development Model—a spiral that contains the phases of assessment, planning that is based on the results of that assessment, implementation of those plans, and then continuing to circle back to reassessment, replanning, and reimplementing. As the organization and its processes mature, these iterations can take place farther apart in time, but, in these times, no organization can ever afford to become complacent (Figure 3.2). Indeed, each time the PMO stretches to take on greater responsibility, a cultural shift will take place, and a new gap analysis will be necessary. Using the processes outlined in this chapter, each new step can, it is hoped, be less traumatic than those that came before.

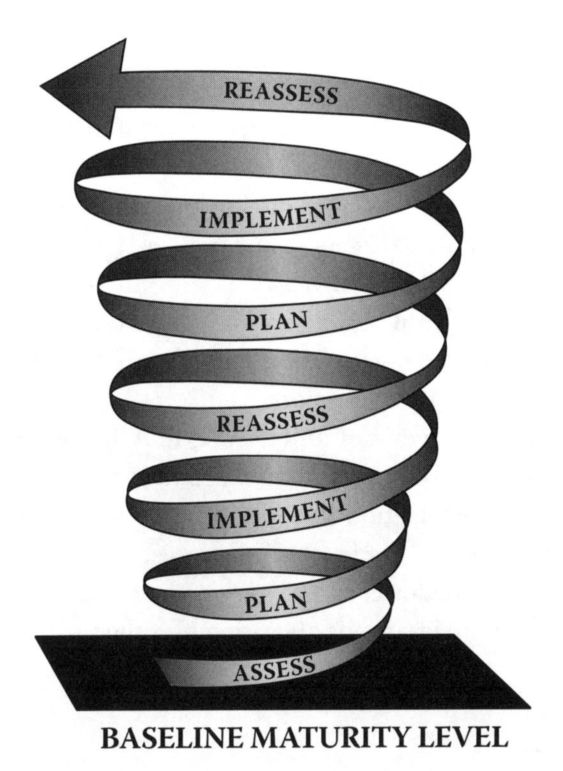

Figure 3.2 Assessment, planning, and implementation iterations.

Talking Points

How the PMO helps organizations boost maturity:

- **Project Support**. The PMO can make the lives of project team members easier by assuming administrative chores in the areas of project scheduling, report production and distribution, operation of project management software, maintenance of the "visibility room," and maintenance of the project workbook.
- **Consulting/Mentoring**. As organizations mature in project management, the PMO satisfies an increasing need for internal project management consultants. These people will provide the organization with the expert insights it needs to execute projects effectively.
- **Processes/Standards**. The PMO is the unit within the organization that develops and promulgates common methodologies and standards relating to project management.

- **Training.** The PMO trains project managers, team members, and clients regarding project management principles, tools, and techniques. Both training material and instructors originate in the PMO.
- **Project Management.** The PMO can house a group of professional project managers who can be assigned to carry out the organization's projects.
- **PM Software Tools.** As the PMO matures, it becomes the focal point in the organization for software tools supporting the project management effort.
- **Portfolio Management and Strategic Alignment.** As the owner of the project portfolio management process, the PMO assists the organization in developing a clear view into the relative values and status of all the projects underway and in the pipeline. It adds clarity to executive decisions about projects and rationality to the allocation of resources.

Notes

1. J. Kent Crawford, *Project Management Maturity Model*, 2nd ed. (Boca Raton, FL: Auerbach Books, 2006).
2. Paul Dinsmore, *Winning in Business with Enterprise Project Management* (West Babylon, NY: AMACOM, 1999).
3. *The Value of Project Management Study* (Glen Mills, PA: PM Solutions' Center for Business Practices, 2000). See also J. Kent Crawford, and James S. Pennypacker, "Why Every 21st Century Company Must Have an Effective Project Management Culture" (CD-ROM presented at the Project Management Institute 2000 Proceedings Symposium, Newtown Square, PA).
4. Software Engineering Institute, *Capability Maturity Model for Software* (Pittsburgh, PA: Carnegie Mellon University, 2008).
5. Software Engineering Institute. *Capability Maturity Model for Software.*
6. Kaplan and Norton, *The Balanced Scorecard* (Cambridge, MA: Harvard Business Press, 1996).
7. Jeannette Cabanis-Brewin, "The Elusive ROI," *PM Network* (April 2000).
8. Interview in *InfoWorld* (March 3, 1997).
9. First described by management theorist Mary Parker Follett in the early 1930s, the Hawthorne Effect refers to the tendency for any workgroup to improve its productivity simply because it is being studied; the extra attention motivates workers to do better than usual.

4

PMO Planning, Preparation, and Strategy

With this chapter, we find ourselves at the threshold of the PMO itself. Before stepping across that threshold, consider how well you have laid the organizational foundations. Like any endeavor, the creation of a PMO requires a strategy, and the key to strategy is to think big, but act in small steps. A Project Management Office (PMO) is a project, and, in the initiation stage, you will need to ask yourself some key, strategic questions.

1. Do I have all the PMO critical success factors in place?
   a. Executive support
   b. Appropriate funding/resources
   c. Acceptance by project managers
   d. Acceptance by business managers
   e. Institutionalized PM culture
2. Have I matched the PMO with our business needs? A PMO that gets too far ahead of, or lags behind, the business needs is doomed. It will be seen as overkill when it has capabilities the business does not value; conversely, it will be perceived as ineffective when it does not have the capabilities to meet business needs.
3. Have I planned for the future by looking for ways to integrate organizational strategy with delivery of projects and programs? You can maximize the value of the PMO by providing improved portfolio governance and balancing, and using project portfolio management to optimize resource management. To successfully make this leap, from project management to managing the organization by projects, you will need to plan to expand the scope of the PMO beyond the usual responsibilities, as shown in Figure 4.1.

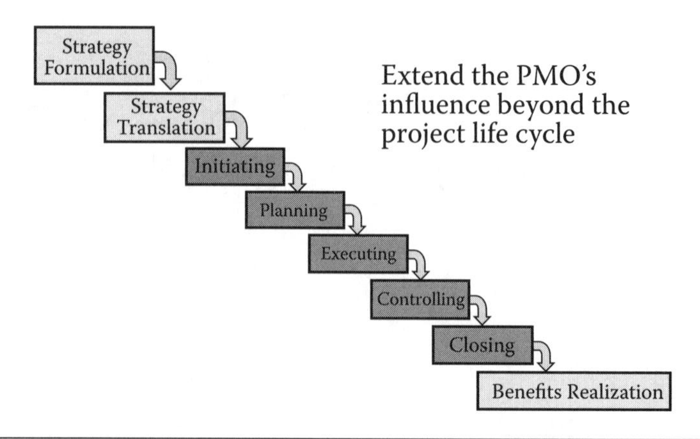

Figure 4.1 Extending the PMO's influence. (From Project Management Solutions, 2008.)

A word of warning: You will not need to study the steps and recommendations here if you have not already dealt with the broader issues raised in Chapters 1 through 3—if you have not made the business case for a PMO, assessed your organization's readiness for it, designed a project office of the appropriate level of integration and sophistication for your particular organization, and received executive commitment for the sweeping organizational changes that will be necessary for success. Without first tackling these critical issues, merely going through the ordinary project management steps for implementation that we are about to address will most likely be in vain. Introducing strategic, enterprisewide project management via a project office to an organization cannot be done from the bottom up in an organization, it cannot be done piecemeal, and it is unfair to the discipline of project management (not to mention the individual project managers) to set your PMO up for failure from the outset by not properly preparing the groundwork.

With that caveat, let's dive into managing the project of creating your project office. As with any project, we'll begin at the beginning: a project charter.

The Project Charter: Agreeing on a Destination

The project charter formally recognizes the existence of a project. It describes the project at a high level and explains the business need for the project. The charter is completed typically by the PMO director or the PMO project manager, and approved by the business executives who are affected by the project, beginning with the project's executive sponsor and other senior managers (members of the PMO steering committee). The project charter authorizes the PMO director to expend company resources in planning the project. With an approved charter, the project is added to the organizational budget. (For help in creating the project charter, see the Table 4.1).

The project sponsor, PMO steering committee, and all stakeholders are identified in the project charter. The charter is coordinated with all stakeholders, who initial the document to indicate their approval. It is then approved by the project sponsor and issued to the project manager.

Objectives and Milestones: The Map to Your Destination

As with any project, planning begins with the establishment of goals, objectives, and milestones. It's incredible how many businesses begin projects without a clear idea of the end, a map detailing how to get there, or a set of criteria to tell them how to know when they have arrived. When you don't define these things very early on in the planning process, your project is almost certain to end up somewhere unforeseen.

The objectives of your PMO initiative should follow the old SMART guidelines: Specific, Measurable, Agreed upon, Realistic, and Time-constrained.

Specific

Whenever possible, the objective should be expressed in terms of defined deliverables. For example, for a project management methodology development project, a specific objective statement might read:

> To achieve a PM HealthCheck average of 3 or better in the organization within a six-month period through implementation of an organizational PMO and acceptance by the PMO steering committee.

Measurable

How will you know when you are done? Your objective should enable you to be clear about when the objective is completed by defining a

Table 4.1 Project Charter Template

PROJECT PARTICIPANTS

PROJECT NAME

Enter a brief name to describe the project. For a project with broad organizational impact, it is wise to choose a name that will generate some excitement about the project.

PROJECT SPONSOR

As discussed in Chapter 3, the Sponsor for a PMO Deployment project must be of sufficiently high level in the organization to influence other senior managers to cooperate with the changes required by the project. This person will be responsible for budgeting the funds to undertake the project and will have final authority to approve project completion.

PMO DIRECTOR/DEPLOYMENT PROJECT MANAGER

The primary liaison with the Project Sponsor and PMO teams. This person is responsible for carrying out all project-related activities.

OTHER PARTICIPANTS

In addition to the sponsoring business area, indicate other business groups that will have crucial responsibilities for the project.

PROJECT DECRIPTION

BUSINESS BACKGROUND

Give an overview of the business reasons for the project. This should include the business justification for the PMO deployment project, value the organization can expect to achieve, why the initiative is being undertaken, and the span of influence of this initiative.

PROJECT SCOPE (DETAIL BOTH WHAT IS IN AND OUT OF SCOPE)

Provide a general description of the project scope (provide details in the following sections). Indicate both what is *within* the anticipated scope and what is *outside* the scope. Consider these topics:

- Systems
- Infrastructure
- Communications
- Business locations

OBJECTIVES

List specific business objectives that the project is anticipated to achieve.

DELIVERABLES

List the specific deliverables expected from the project and how these will fulfill the objectives. The deliverables should be as tangible as possible. See the discussion in this chapter on Phase II and Phase III for some examples of deliverables.

CONSTRAINTS

List factors that will limit the project team's options. For example, a predefined budget range is a constraint that is very likely to limit the team's options regarding project scope and staffing levels.

ASSUMPTIONS

List factors or situations you will assume for the purposes of planning the project. For example, if the availability date of a key resource is uncertain, the team should make a reasonable assumption about the date of availability and list this as an assumed or contingent factor in the plan.

We agree that this is a viable project. We authorize the beginning of the Planning Process.

Table 4.1 Project Charter Template (Continued)

DATE:
Project Sponsor
Member of the PMO Steering Committee:
Member of the PMO Steering Committee:
Member of the PMO Steering Committee:

measure or set of measures. In the example above, the project will be finished in six months upon approval of the steering committee.

Agreed Upon

All the stakeholders should buy into the project objective; in most cases, this means senior management across several divisions, but could include vendors, suppliers, and other industry partners.

Realistic

The objective should also be realistic in that it is obtainable with the time and resources allocated. Too many PMO initiatives reflect an unreachable objective that isn't obtainable even with unlimited resources and unlimited time, or which, even if obtained, cannot be objectively measured. Referring to our example, before the time and cost objectives were set, the project manager may have issued a "request for Information" to several consulting companies to ensure the objective was realistic and attainable.

Time-Constrained

Project management is all about meeting deadlines. In the case of a PMO, the objectives should be staggered in phases to allow the project to both meet immediate project needs and address longer-term issues involved in changing the organizational culture to a project-based one. Notice in the example that part of the objective is a firm completion date in six months. The following interim milestones also may appear in the charter for this project:

- Gather current PM practices being used within the company in one month
- Kickoff meeting with outside consultant in six weeks
- Conduct *PM for executives* workshop in two months
- Draft of *initiating section* in ten weeks
- Draft of *planning section* in three months
- Draft of *executing section* and *control section* in 14 weeks
- Draft of *closing section* in four months
- PM *methodology workshop* with cross-sectional team for review and refinement in 18 weeks
- Revisions completed in five months
- Project completed in six months

Using Gap Analysis to Set Milestones

The PMO project charter will ultimately include the project objective statement. Results from the gap analysis are incorporated into the project objective statement, project milestones, and the detailed project plan (see Table 3.6 in Chapter 3, the sample gap analysis chart). Gaps identified are incorporated into the project plan as specific outcomes to be achieved. This book does not cover basic project planning, but suffice it to say that the planning process is carried out in the same manner used for any project. Keep in mind the objective of this particular project is to deploy a project office; therefore, the project objective may be a little broader in scope than the typical project objective.

A note of caution: Many times companies skip this step and approach the PMO design by merely looking over an organizational chart and identifying which functions they want to transfer into the PMO. In our view, this approach merely skims the surface of the value-adding change that a PMO, especially a Type 3 Strategic PMO, can bring. It is prudent to concern yourself with function identification, not just function transfer. Future needs, while perhaps beyond the scope of the initial deployment, should become long-term goals for the next three to five years.

A PMO should be established based on an organization's needs, both short term and long term. Specific project objectives and milestones flow from those identified needs, and the staffing plan is determined based on those objectives.

Delivering Value with Specific Short- and Long-Term Objectives

Where to start? The best way to win converts for the PMO method of managing projects is by adding value and getting results as quickly as you possibly can. Even senior management executives who are receptive to implementing a PMO have probably been involved far too many times with programs that took two to three years of implementation before any results were shown. Money is tight, competition is fierce, and companies have raised the bar on expectations for major change initiatives. It's likely that your management will be looking for fairly immediate results out of a PMO deployment. To satisfy these concerns, find a set of short-term objectives that provide immediate value, such as developing, deploying, and supporting a project management methodology that brings immediate improvement to one or two pilot projects. At the same time, continue to work toward longer-term objectives related to changing organizational culture and adapting the organization to a new way of doing business. If you don't add value quickly, you will lose momentum; however, if the larger-scope organizational change agenda is not pressed, the organization won't derive the greatest benefits from the project. Again, the gap analysis should help bring to the surface the areas in which the problems with projects are most pressing.

Specific long-term objectives should focus the organization on achieving increasing levels of the Project Management Maturity Model (PMMM), enabling optimized project management performance as an end result. Long-term objectives might include:

- Achieve Level 2 of the PMMM in six months
- Achieve Level 3 of the PMMM in one year
- Achieve Level 4 of the PMMM in two years
- Achieve Level 5 of the PMMM in three years

Implementation Strategy

The PMO approach should take a tactical focus in the beginning considering immediate concerns, business necessities, and the minimum requirements necessary to jumpstart the change process. During this initial period, preliminary steps can be undertaken to lay the groundwork on broader, complex issues. Long-term solutions address

Table 4.2 Short- and Long-Term Objectives for the Project Office

PHASE I: ESTABLISH PROJECT OFFICE

- Identify and prioritize all projects
- Deploy Project Management Office methods
- Train core teams
- Successfully complete pilot projects
- Attain management oversight on pilot projects
- Establish time and cost collection by project

PHASE II: STARTUP WITH SHORT-TERM INITIATIVES

- Train all project teams
- Utilize project management methods on all projects
- Plan, track, and manage resources
- Collect and manage projects
- Establish the project management costs for all culture
- Integrate management oversight into all projects
- Implement project reviews and audits

PHASE III: ROLLOUT WITH LONG-TERM SOLUTIONS

- Train all business teams
- Fully integrate PM throughout organization
- Integrate resource and cost management across the organization
- Keep management actively involved utilizing PMO reporting and analysis

PHASE IV: SUPPORT AND IMPROVEMENT

- Implement a Continuous Quality Improvement Program

permanent maturity efforts that result in long-term value to the organization and ensure that you achieve your time-to-market timeframe. To accomplish this, we have found that a four-phase approach works well (see Table 4.2).

Phase I: Establish the Foundation

In this phase, define the PMO and determine your immediate concerns and long-term objectives. As appropriate, start with an assessment of your current capabilities, goals, and objectives. Baselining against the PMMM identifies the baseline positioning of project management within the organization and aids in planning future tasks and activities. A series of meetings is held with key stakeholders and subject matter experts to understand current capabilities, challenges, issues, and goals.

Based upon the discussions, an assessment report is developed that captures the current state and future vision along with an improvement plan recommending short-term initiatives and long-term solutions.

After developing the top-level improvement plan, determine the PMO functions and staffing, identify stakeholders (to include key management, mentor programs, and pilot projects), and prepare a communications strategy. This phase ends with issuance of the project charter, authorizing the PMO project team to proceed with funding and fill immediate staffing needs. The timeframe for this phase varies widely, depending on the organization. Some companies feel driven to ramp up their PMO quickly and can lay this organizational groundwork in a matter of a few weeks. Others will spend a longer period on the assessment process. The determining factor is the urgency of PM improvement.

Phase II: Startup with Short-Term Initiatives

In this phase, we start up the PMO, put in place short-term initiatives, and initiate the project mentoring effort. PMO startup includes staffing the office for near-term needs, initiating communication activities, and making the organization aware of the PMO and its responsibilities.

Two efforts are initiated to demonstrate the immediate value of the PMO within the organization: short-term initiatives and project mentoring. The short-term initiatives provide solutions to immediate concerns and take care of issues surfaced by key stakeholders, solutions that can be implemented quickly.

Examples of short-term initiatives include:

- Deployment of a project management methodology (see Chapter 5 for detailed information on methodology and standards).
- Building an inventory of your projects (new product development, information technology, business enhancements, etc.) as a basis for project portfolio management.
- Preparing an executive report, showing the status of all active projects.
- Establishing summary project report structures and project success metrics.

- Organizing brown bag training lunches: Brief, informal training sessions that familiarize members of the organization not only with the PMO initiative, but also delivering key project management concepts.
- Establishing support for new projects and projects in need.
- Conducting project planning or project control workshops.
- Identifying and deploying one or more pilot project initiatives.
- Providing templates for recurring project activities (see Chapter 5 for a discussion of using templates to encourage adherence to project management standards).

In conjunction with these short-term initiatives, project mentoring can be kicked off almost immediately, either by using the experienced, previously successful project managers already on board or by soliciting the assistance of external consultants experienced in the mentoring process. Project mentoring is an excellent way to provide immediate project management value to projects that are in the initial startup phase or are in need of support without waiting for the implementation of formal training programs or process rollouts. For more information, a discussion of mentoring is included in Chapter 7.

Phase II ends when the short-term initiatives are in place and the team is ready to focus exclusively on the longer-term solutions planned in Phase I. In addition to the initiatives discussed, Phase II may include:

- Training the core teams—teams for the pilot projects and for projects associated with PMO startup, such as methodology development.
- Beginning to involve management through the oversight committee review meetings.
- Beginning to collect time and cost information by project in order to do the project tracking that will validate benefits to the organization.
- Implementing an active communication plan to secure the confidence of the organization that these projects are going well and also that they have control and oversight on the pilot initiatives by communicating quantifiable results on the pilot projects. This will gain positive recognition from the business units because they are used to seeing projects fail or struggle.

- Developing and deploying processes and standards (see Chapter 5).
- Establishing a bench of project managers.

Phase III: Rollout with Long-Term Solutions

There are increasing benefits to an organization as its project management capabilities mature. Phase III focuses on improving/streamlining the processes, developing personnel, and putting in place the more permanent support structure necessary for project management to succeed. In this phase, we develop the long-term solutions, continue the project mentoring effort, conduct additional pilot tests (as appropriate), and gradually roll out the fully functioning PMO. Examples of critical success factors include:

- Continuing development and tailoring of processes and methodology
- Development of a training curriculum
- Development of detailed reports and metrics
- Addressing resource management issues
- Tool deployment (see Chapter 8)
- Project portfolio management (see Chapter 6)
- Project manager career progression and certification (see Chapter 7)
- Organizational change management and transition planning (see Chapter 9)

All of these items take time to develop, and the deployment should be done incrementally starting with pilot tests on selected projects. The assessment and improvement plan completed during Phase I provides the overall long-term goals and objectives for the PMO, and this phase develops, pilot tests, and rolls out the methods, standards, training, and support activities to achieve those overall goals. Other Phase III activities not mentioned may include:

- Training all the project teams
- Implementing the methods established in Phase II for planning, tracking, and control
- Integrating projects into programs

- Implementing reviews and audits
- Establishing competency standards for project managers

We move yet to another level of maturity when we begin to integrate project management throughout the organization. By the end of Phase III, all project-related estimating, budgeting, scheduling, change control, variance analysis, time tracking, issues tracking, risk analysis, and project reporting should be carried out under the auspices of the PMO.

Phase IV: Support and Improvement

In this phase, the PMO is in full operation and is supporting the organization's projects both from a tactical and strategic perspective. The PMO conducts day-to-day activities, refines project management activities, and expands the involvement of the PMO where appropriate. Training and other initiatives continue under the direction of the PMO. Key stakeholders provide feedback on the PMO's efforts, and activities are continually refined as part of a quality management program. Portfolio management becomes more sophisticated as more project metrics are collected. A lessons learned library, benchmarking, collecting best practices, and other knowledge management activities are hallmarks of the mature Strategic PMO in Phase IV.

From our consulting practice, we have gathered the following ten practical keys to the successful deployment of a project office.

A Project Office or *a Project Management Culture?*

Is it possible to effectively deploy a PMO without changing the organization's culture to a project management way of doing business? I contend it is not. To be effective as a PMO (if it is to enable significant improvements in the ways projects are managed) requires an organization to mold itself into projectized form. Effective project management must become the core of how all projects are conducted. The PMO will be ineffective if its only role is generating timelines and reports without the cultural change in how projects are initiated, planned, executed, controlled, and closed out. Movement to a project management culture throughout the organization is critical to success in managing the company's projects.

Ten keys to success for deploying the project management culture in an organization include:

1. **Keep it simple**. First and foremost, be realistic and work the basics. If your staff can't explain why they are doing a particular project and they can't identify their 60-day plan, focus on helping those areas first. Don't worry about a sophisticated estimating process yet; focus on simply understanding project goals and developing basic plans. Once you identify these basic needs, stay focused and don't do too much too soon. Employ the minimum project management essentials (such as a project charter, project management plans, project schedules, project metrics, and project reporting) and start up the office to help project teams. Don't try to optimize every aspect of project management.
2. **Communicate**. The best idea goes nowhere if you keep it to yourself, surprise everyone at the last minute, and expect it to be accepted and practiced. People don't like surprises, so explain what you are doing and why, frequently and in plain English. This is one of the keys to successfully creating a project culture, discussed further in Chapter 9. Let everyone know how the PMO and the new business practices will help them. Package a "story" and spread it around. Say the same message over and over, tailoring it for the different levels of the organization. Communicate your goals and successes via different avenues: a project bulletin board, status review meetings, brown bag sessions, e-mail, or communiqués. Just get the word out.
3. **Make sure that expectations and goals are shared**. Make sure the charter for the project office deployment project is endorsed by all stakeholders. Have a kickoff meeting—a big event—to share the elements of the charter, the goals, and vision of the executive sponsor. Have the sponsor say a few words, focusing on the benefits to be achieved. Keep people informed as you create the project office. Make a big deal out of your successes.
4. **Focus on value**. Determine the organization's most pressing concern and fix it. Find what hurts the most and focus on it. Talk to key stakeholders at all levels within the organization.

Try to fix one key concern for each level. Sometimes the immediate fix is an interim solution that is done inefficiently (such as manual reports), but at least the report provides information and insight with some degree of confidence. Whatever you choose to do, link the goals of the PMO to the organization's goals and explain how the office and project management practices help meet the organization's goals. Immediate results in selected areas are important to keep interest and excitement about the PMO churning as well as to prove to executive management that the PMO isn't just business as usual. Be very clear in identifying the deliverables at select phases of the pilot projects to show that results are occurring.

5. **Support project managers**. Often someone who has been a wonderful technician or a proficient business analyst or engineer is placed into the role of project manager with no training, no assistance, no support; then we wonder why they struggle. A key to success is providing support, assistance, mentoring, and guidance to project managers. They need support to help develop the plans, manage the schedules, monitor the costs, manage the resources, do the variance analyses, and generate the reports. When a project manager is expected to do all the specialized work on a major project, he or she cannot focus on the areas where project managers add the most value—the "art" of project management, which involves communication, facilitation, negotiation, creative problem solving, and other critical tasks.
6. **Take time to understand the organizational problems from various points of view**. Project managers don't just deal with the executive level or only with their project team members. In order to gain widespread acceptance of the PMO throughout the organization, the PMO director should take time to learn about the issues and challenges facing all the departments or business units that will be affected by the changes. In particular, the input of the technical staff should be included in the project plan because this can be crucial to issues of change and risk as well as configuration management. The PMO director will be able to work much closer to the plan if the technical risks have been identified, assessed, and planned for.

7. **Conduct pilot tests**. No two organizations are the same. There are different organizational cultures, personalities, approaches, techniques, and technologies. This uniqueness requires us to begin implementing a PMO by conducting pilot projects, deploying methodology and process against those pilots, and then refining the processes and methodologies with lessons learned for subsequent deployment. Through pilot testing of the enterprise project management approach, we are able to gain experience, adapt and reapply lessons learned, and be much more successful in the enterprise deployment of PMO.
8. **Establish incremental goals**. Research on project failure[1] tells us that in order to be successful, projects must be broken down into phases or periodic review stages. This applies to any project undertaken in the organization, not just the project of implementing a Project Management Office. At the end of each phase, we can look at where we are, look at where we have been, compare our progress to where we need to be going, and revaluate our approach, redirect our efforts, reprioritize our initiatives, and reestablish our commitment for management that the projects are valid and critically important. At one time, midstream corrections of this type would have been frowned on in project management circles. Thankfully, however, the discipline has established the foolhardiness of holding fast to a course of action when the surrounding environment is changing.
9. **Involve the right people up front, starting with your executive sponsorship**. No matter what you do, without executive sponsorship, you will fail. Make sure you understand who cares, who will be impacted, and who makes decisions. Get the leadership team involved from the beginning. Find out their needs, expectations, and goals. Identify their concerns and work to address them. Remember to keep it simple, focus on value, and plan. Understand the problems at different levels. Identify an executive "cheerleader" (or sponsor) and encourage as much "cheering" as possible. Plan regular status review meetings with the PMO steering committee. But, don't focus only on the executives. It's important to get the right people on the project teams as well. Don't choose these important pilot project teams strictly because they were

available. Involve people who are most knowledgeable in the technology, the process, and the business area. By getting the right people involved in the initial planning stages, even if you have to delay the initiation of planning for the project, you can do a much more effective job of planning, identifying the issues, identifying the risks, and planning for success for the future. The "right people," it should be noted, also may include stakeholders from outside of the immediate organization, such as clients or vendors. If implementing new technology is to play a major role in your PMO initiative, it only makes sense to include a knowledgeable vendor, rather than relearn all those skills or make all the mistakes that a more experienced person might be able to foresee.

10. **Plan.** We've covered planning in considerable detail already, but it is one of the real keys to success. Although it is sometimes painful and may at times appear to be nonproductive, take the time to plan thoroughly up front. The plan will help set expectations and facilitate communications. Establish incremental goals to show progress and results to the organization. Identify specific short-term and long-term solutions and explain how, in some cases, an interim solution will set the stage for a long-term objective, e.g., a current report that is done manually may need to be automated. Make sure you plan enough time to conduct pilot tests and train individuals before setting in place the new process or tool. So many times we launch into a frenzied work effort because of the attention the project is receiving from management. We want to show results immediately, but often in doing so we begin working away at all the wrong things. It may not be necessary to plan the full PMO project in detail. Rolling wave planning allows us to fully plan out the first phase of the project and then, at a higher level, plan out the remaining phases. This allows us to incorporate lessons learned from earlier phases into subsequent phases of the project.

Related to this item is to *take time to adequately train the project teams.* The great majority of teams are not prepared to embark on this new discipline. Many don't share a common terminology. Many don't understand the techniques. Many don't understand the business case, or the "soft" side of project management. And many don't appreciate the rigor that is necessary for effective planning. Most of the time, project teams are simply anxious just to get started doing the work. Unfortunately, although many project teams are absolutely convinced that they are on the right track, past rates of project failure tell us that it's likely project teams will hit a roadblock and find out that they were going down the wrong path all along, having wasted time, resources, and money along the way. We now know that many project derailments can be avoided with appropriate planning, forethought, risk identification, risk analysis, and workaround considerations. Therefore, training to transfer knowledge of project management process is absolutely critical to get everybody working on the same wavelength as well as to prevent resistance to the organizational changes wrought by the implementation of a PMO.

However ... Five Ways to Fail

Just as there are key activities that work in a PMO implementation, there are factors that hinder progress. Avoid doing these things. At a minimum, recognize what is happening that may require a change in behavior and approach. Your implementation will fail if you:

1. **Forget key stakeholders**. Earlier, we mentioned the importance of executive sponsorship. However, executives are not the only key stakeholder or customer of the PMO. Others include project managers, project teams, functional/resource managers, and line managers. Just like the executives, these stakeholders must be involved from the beginning. Determine their needs, expectations, and goals. Understand the problems from the executive's point of view; otherwise, you may overlook a key concern. Even if buy-in has been developed and a blessing has been bestowed on the project charter, it's important not to let communication with stakeholders diminish as the project progresses.
2. **Demand before providing**. A PMO must be viewed as an entity that helps, an entity that *provides* services to ease project management administration and to facilitate smart business practices. All of this results in an improved track record

of project delivery. The PMO should never be in a position of always demanding information and seldom providing services. You will not be successful by asking for too much too soon.

3. **Do it all at once**. There are three factors to a PMO implementation: people, process, and tools. Obviously changing all three at once is a very complex undertaking. If possible, avoid doing this. Change the environment (tool), but keep the process the same, or change the process, but use the same environment and tool. A phased approach makes this feasible. As PMO director, don't do it all at once. You may not be able to deliver, and people will get confused. Don't allow over eagerness on the part of executive sponsors to push you into making promises you cannot keep or that will overextend the resources at your command. It's better to succeed incrementally than fail spectacularly.
4. **Procrastinate.** Once a decision has been made to implement a Project Management Office, move on it. Don't hesitate or partially support the idea. You will lose support and focus. The organization will stop believing in the concept. In addition, the longer it takes to implement, organizational changes and upheavals may occur to disrupt the PMO initiative. Such adjustments may result in changes in executive sponsorship and other key stakeholders. Priorities may change and the effort may lose support and funding, resulting in a failed initiative. When implementation gets prolonged due to administrative issues or organizational restructuring, decisions are postponed on the budget, administrative support is not given until a crisis occurs, staff is pulled off to work on other projects, etc., and soon the PMO initiative is moribund. Plan thoroughly—and hit the ground running.
5. **Work in a vacuum**. In a PMO implementation, a team approach wins. The office is intended to serve multiple customers, each of whom have personal experiences and ideas to share. Incorporate other people's ideas and acknowledge them and give credit where due. Learn from others' experiences; don't reinvent the wheel. Find out individual requirements and needs, and design accordingly and appropriately. Leverage all the knowledge and experience at your disposal.

Measuring Success: How to Know When You Have Arrived

As we've detailed, the key to deploying a PMO is to make sure that there is added value to the organization, that the value accrues incrementally, and that the value is communicated to all stakeholders. Therefore, two things need to be in place: (1) metrics to define success at each stage of the deployment and (2) a communication plan to get the good news out.

Metrics

"When performance is measured, performance improves. When performance is measured and reported back, the rate of improvement accelerates."[2] While project management metrics can be a confusing subject, when implementing a PMO, you are fortunate to have the one thing that must be in place in order for measurement to be meaningful: a baseline.

The baseline in this case is provided by the assessment of organizational maturity done as a necessary first step (see Chapter 3). By reassessing the organization using the same assessment tool and maturity modeling approach that you started out with, organizational improvements in project management can easily be evaluated.

As for measuring progress on the PMO deployment project itself, we can use the tried-and-true metrics we use for any project: variance analysis. Once the charter is issued, the team creates the work breakdown structure (WBS). Then, using the WBS, the project schedule is created and displayed either as a logic network (precedence diagram/critical path network) or a Gantt chart with relationships among tasks embedded. By conducting the usual periodic progress meetings and updating the Gantt chart, schedule variance can immediately be determined and acted upon. Cost variance is not difficult to measure in a project such as this for the simple reason that most of the expense is either in payroll of internal team members or contract costs for external consultants. Identify cost variance and take corrective action. Finally, technical variance and scope change must be addressed. Using the list of desired capabilities in the short, mid, and long term discussed earlier, create a requirements matrix to ensure that each desired capability is addressed by the WBS and project plan. Create milestones for each capability and track progress toward those goals. Identify variances and act promptly to keep technical progress on track.

One important step when measuring the success of the implementation is to revisit the gap analysis and do a "reality check." Have you covered all the areas of weakness that were identified? Have the issues raised been addressed? And, finally, simple though it seems, is everybody happy? Simply surveying all the stakeholders to get their input on how they feel the new initiative is working can be extremely valuable.

Communications Planning

As part of the initial project planning effort for the PMO initiative, a communications plan must be in place. First, there will be the need to communicate about the initiative to the overall organization. Then there will be specific planning for meetings related to assessment and deployment.

Early on in the pilot stage, there should be opportunities to generate success stories. Rather than waiting to stumble over these opportunities, however, they need to be programmed into the deployment and integrated into the communication plan so that, as they are experienced, they are communicated back to the appropriate stakeholders: management.

The Meeting Guide (Table 4.4) describes a variety of forums that can be used to communicate throughout the organization and through all project phases. Little progress can be made in an organization without effective communication. Table 4.4 includes some communications ideas to consider in deploying for your organization.

Keeping everyone informed without swamping them with an overload of information is key to creating a sense of "ownership" at all levels of the organization (see also Table 4.3).

Best Practice: Communications

A large insurance company, during their project office deployment, had a wall outside the cafeteria where anyone with any concerns or issues or questions about the Enterprise PMO deployment could just stick a Post-it® note up on the wall. Responses to questions and issues

Table 4.3 Project Communications Plan Form Definitions

FIELD	INSTRUCTIONS
Audience (Who needs to know?)	Who is the audience for each communication? Check the Project Charter, Statement of Work, and other project documents to determine audiences. Some messages will go to audiences defined by function or group membership: • Project (key project stakeholders, project personnel, project managers, project sponsors, business area project manager, consultants) • Business Area (business group participants not on the project team, cross-business groups, business group by business group notification) • Corporate (Executive Committee, selected executive officers) • Outside customers (customers who use the project) Some audiences will be defined by project phase, milestones, and status: • Introductory audience • Audience for various phases and milestones • Testing audience • Implementation audience, by phase • Conclusion audience for project review and sharing the success
Message (What?)	Describe the message that needs to go out to this audience: • What does the project need to communicate to its audiences? • Who is authoring, sponsoring, and/or standing behind the message? • What's going to happen? What other needs or work is it related to? • How far along are we? When is it going to happen? • Where's it going to happen? Where's it *not* going to happen? • How is it going to take place, in what steps or increments? • How will the project team help you get through the change? • What does the recipient need to do, and by what date? • When will there be further communications, second warnings, etc.? • Where can they get more information, who should they call?
Intent (Why?)	Why is this communication taking place? • What is the intended effect? What do we hope to achieve? • What are the benefits?

(*continued*)

Table 4.3 Project Communications Plan Form Definitions (Continued)

Media (How?)	*How* to communicate will depend on the phase of the project, the audience, etc. It generally takes face-to-face communication to achieve buy-in, support, and to get someone to take action. At other times, you will use hard copy print and electronic media, or combinations of media.
(When?)	Consider the Statement of Work, the evolving project plan and the advice of project leaders and key stakeholders to determine a communication approach and timing.
Responsibilities	For each message in the Project Communications Plan: • Who will prepare the message, develop the media, and coordinate the delivery? • Who will author or sign the communication? (Who is the message from?)

also were posted. As people were walking to and from the cafeteria, they would stop and see what the latest issues were, the latest concerns, the latest feedback. And many of those questions found their way into the regular newsletter that was published, and those in charge of the PMO responded and addressed the issues there. In many cases, the CIO or one of the executive VPs would respond and, in responding, show their support and encouragement, show insight into why the PMO project was valuable, and why everyone should support it.

Purpose

Communication among the various entities working on (or interested in) a project is absolutely essential. The purpose of a formal

Table 4.4 The Meeting Guide

DESCRIPTION	PURPOSE	AUDIENCE	MEDIA
One-on-one	Address "What's in it for me"	Sponsors/management	Face-to-face
Kickoff meeting	Inform	PMs and sponsors	One to many meetings
General information	Information updates	Entire company	e-mail, intranet, etc.
Intranet BBS	Questions, comments	Entire company	Intranet
Newsletter articles	Information updates	Entire company	Electronic or paper, but more formal than e-mail or memos
Team meetings	Status	Team	Face-to-face

Project Communications Plan is to ensure that all teams and interested parties who are involved in any way provide and receive appropriate communications.

The involvement of multiple teams and organizational units enlarges the web of necessary communications and increases the complexity of conveying the right message to the right audience at the right time.

Key reasons for project communications include:

- Establishing (and maintaining) the support of those involved, including project sponsors, team members, and those who will use the project deliverables.
- Educating decision makers on the "whats" and "whys" of the project.
- Informing the ultimate beneficiaries and others who will be affected by the project and preparing them for what to expect.

In addition to the effect on those closest to the project, other impacts include:

- Any phase of project implementation (e.g., testing) may involve some change or even disruption in regular services to some or all users. These impacts need to be communicated in advance.
- Implementation also may involve changes in local procedures, new training for users, and other effects. All possible impacts need to be communicated in a timely manner.

Origination and Timing

The project manager completes the Project Communications Plan in consultation with the business area project manager and other key project participants. Much of the requisite information for the communications plan will come from the statement of work (SOW), so it is suggested that the Project Communications Plan be completed *after* laying the foundations in the SOW. Then, the Project Communications Plan can be summarized in the body of the SOW and appended in its entirety to the SOW.

Inputs

- **Statement of Work**. Even if it is not 100 percent completed, the SOW is a reference for communications planning.

Outputs

- **Project Communications Plan**. Use the form to organize your plan. Changes and updates to the completed Project Communications Plan should be stored in your central project files.

Other Outputs

- **Tasks.** As you prepare the Project Communications Plan, you will identify additional tasks or task groups to include in later versions of the Project Plan.
- **Issues.** Issues that arise during preparation of the Project Communications Plan should be logged into the project issues log.

Project Communications Plan Guidelines

Communication is not a single event, it is a composite, the result of several messages that build on one another. In your Project Communications Plan:

- Strive for "no surprises." The objective is that everyone who needs information about any aspect or phase of the project gets the information they need in time to assimilate it and, if necessary, respond to it.
- Communicate "down the chain," that is, communicate with leadership first, then team members, and so on.
- Define your audiences carefully—who needs to know and who does not. You will have different audiences for different aspects of the project.
- Anticipate the information needs of each audience and time your messages to coincide with project milestones and the audience's anticipated need to know.

- Target each audience with appropriate media—different audiences may require different approaches.
- Plan for multiple messages to the same audience, with repetition and reinforcement. Repetition will not necessarily upset the recipients, as long as you target them well and send the right message.
- Create a mechanism for anonymous feedback from your audiences.

PMO Value-Adding Strategy: Rein in Runaway Projects

The improvements that have been made in project failure rates over the past few years are in large part due to organizations learning how to control, even to stop, projects that are in trouble.[3] Virtually every organization, no matter how mature and experienced, will occasionally be forced to deal with one or more failing project. Rather than ignoring the problems, it's useful to view the struggling project as an opportunity to learn and input for continuous process improvement. How do you know when a project is in trouble?

Early Warnings

Although it is customary to focus on schedule and budget overruns as red flags of troubled projects, these indicators come too late in a project to forestall problems. According to Gartner, Inc., there are four early indicators of runaways that can often be identified as early as the planning stage so that potential problems can be nipped in the bud.[4]

1. **Inadequate project planning**. This includes ambiguous milestones, failure to chart the interdependencies between tasks and establish a critical path, estimates that are not based on actual organizational history and experience (lessons learned), and a lack of attention to cross-project resource dependencies (in a multiproject setting). Gartner, Inc. has noted that as the number of projects in an organization increases, resource scheduling becomes more important because resource management errors, even on small projects, have a cascading impact on the crowded organizational resource calendar.[5]

2. **Poorly defined objectives and requirements**. When the business objectives of a project are muddled, or when the customer (end user) has not been involved in defining the requirements for a project, the project is likely to go off track or to deliver an end product that is not satisfactory to the customer. Worse, a project may "succeed" in terms of meeting time and cost constraints, but be a failure because it does not answer the business needs of the organization.
3. **Technology disconnects**. The project whose schedule depends on estimated productivity gains from the use of new technology or tools is likely to derail. The effort required to implement complex new tools cuts into productivity; in addition, the change management issues surrounding the introduction of new technology can create cultural upheavals with no quick and easy "fix."
4. **Missing skills**. Failure to assign or identify all the critical skills when doing resource allocation. Whether this problem is later addressed by seeking help from outside consultants or by acquiring new skills through training, the schedule and budget will be significantly impacted. "No project should be undertaken unless the required skills will be available on an as-needed basis," says Gartner.[6]

Another early-warning signal is extreme project complexity. The Standish Group has found that breaking large, complex projects down into component projects—with smaller teams and shorter timeframes—is a key success factor.[7]

In-Progress Problem Indicators

- **Schedule and budget overruns caused by restarts and rework and scope creep**. Schedule slippage of over 10 percent for any task, or a scope creep (the accretion of additional requirements once a project is planned and under way) of more than 5 percent, is generally a serious red flag. Restarts and rework are classic runaway symptoms. In 1994, the Standish Group found that for every 100 failed projects, there were 94 restarts. Each time significant rework or a restart occurs, the team must revalidate the project's scope and objectives, reestimate, reevaluate risks, and decide whether to proceed. Be on the lookout for faulty task management, leading to deliverables that are consistently late and abrupt scheduling changes.

- **Poor communications**. Reports that are inadequate (or overwhelming); insufficient documentation, ineffective meetings.
- **Team dysfunction**. This can be demonstrated by low morale and high turnover, by team conflict, or by confusion and disorganization. Don't gloss over conflict by dismissing it as "personality clashes," it can sometimes reflect deeper problems. Team members on a failing project may develop a "bunker mentality," treating anyone who points out problems as a "troublemaker." When the team is constantly requesting more details or bogged down in meetings, this may call for intervention.

Pulling on the Reins

Where should the PMO start once the runaway and failing projects have been identified? First, *focus only on the highest priority projects*. These can be identified by examining the business case and identifying the tangible and intangible benefits to be derived from project completion. This is a process the organization must master anyway in order to succeed at project portfolio management, and the selection and prioritization of projects is a key function of Strategic PMOs. Once the highest priority projects have been identified, initiate a project recovery program:

1. **Assess the problems**. Identify the specific problem or problems that were the root cause or causes of the loss of control. Look especially hard at the "early warning signals" for clues.
2. **Develop an action plan to address them**. Even if the project cannot be saved in its current form, its likely that some components of it may be restructured into manageable phases. However, don't restart without a strategy in place to prevent recurrence of the problem on the current project as well as on future similar projects. For example, if missing skills were the cause, implement an appropriate training program.

3. **Get commitment to work the action plan from the stakeholders**. This includes team members, executive management, and other stakeholders, including suppliers and subcontractors. The inclusion of all stakeholders cannot be overstressed. A project cannot be successful within the context of an organization whose long-term business needs are not being met. Thus, stakeholders external to the project—subcontractors, users, government agencies, and the like—should have a significant interest in and focus on how projects are managed.
4. **Set project standards**. *The Guide to the Project Management Body of Knowledge (PMBOK®)* provides the most widely known standard for managing individual projects. Update project plan templates and methodology documentation to capture the results of the analysis phase and to ensure mistakes are not repeated on future projects.
5. **Provide coaching and mentoring to the project staff**. According to a report in *Computerworld*,[8] training in the "soft skills" (including leadership) is perhaps the most important factor in improving project performance.
6. **Be sure knowledge transfer is taking place during the recovery**. Learn from your failures.
7. **Stay on top of it**. Conduct reinforcement reviews.[9]

A decision to implement a PMO does not need to be followed by a lengthy, drawn-out implementation. You can't afford it; there is little time, limited resources, competitive pressure, and the need to do business. The way to implement a PMO is to first focus on immediate value and business necessities. You should design an implementation approach that takes care of these immediate concerns and, in parallel, lays the groundwork for longer-term solutions. The key is to keep the implementation simple, focused on value, and structured with a plan. Don't try to do it all at once. Build an office that provides services to ease administration and put in place smart business practices. The net result will be a structured, consistent method to manage projects and an understanding of project performance, resulting in overall better project performance.

Notes

1. The Standish Group, *The Chaos Report* (Boston, MA, 1999).
2. Thomas S. Monson, *Pathways to Perfection* (Salt Lake City, UT: Deseret Book Company, 1996).
3. Jeannette Cabanis, Interview with Jim Johnson, *PM Network* (September 1998).
4. D. Brown, R. Hunter, "Putting the Shrapnel Back in the Grenade: Recapturing the Runaway RAD Project," Gartner, Inc., *Strategic Analysis Report* (July 5, 1996).
5. Brown and Hunter, "Putting the Shrapnel."
6. Ibid.
7. Jim Johnson, "Turning CHAOS into SUCCESS," *Software* (December 1999).
8. Julia King, "IS Reins in Runaway Projects: Users Fight Failures with Better Management," *Computerworld* (February 24, 1997).
9. Richard W. Bailey II, "Six Steps to Project Recovery," *PM Network* (May 2000): 33–34, 36; Paula Jacobs, "Recovering from Project Failure," *Infoworld* (September 27, 1999); Lauren Gibbons Paul, "Turning Failure Into Success: Maintain Momentum," *Network World* (November 22, 1999); and J. Roberts and J. Furlonger, "Successful IS Project Management," Gartner, Inc., *Strategic Analysis Report* (April 18, 2000).

5

Establishing a Project Management Methodology and PMO Governance

Methodology is one of those fine-sounding words that, many times, is used in such a way that it delivers less than it promises. Even the *American Heritage Dictionary* notes that, in recent years, the word *methodology* has become merely a pretentious synonym for *method*. Because one of our editors has pointed out that imprecise use of words often stems from fuzzy thinking about the subject they describe, it seems worthwhile to begin this chapter by defining our terms.

We have repeated throughout this book that one of the primary functions of a Project Management Office (PMO), and a source of almost immediate value, is the promulgation of a methodology: the establishment of methods, standards, and processes.[1] What do each of these words really mean and how do they interrelate?

Defining Our Terms

A methodology is "a body of practices, procedures, and rules used by those who work in a discipline or engage in an inquiry; a set of working methods.[2] But it's also more than merely a collection of methods, it's a framework for making sense of them. As the *American Heritage Dictionary* goes on to say, there is "an important conceptual distinction between the tools of scientific investigation [methods] and the *principles* that determine how such tools are deployed and interpreted."

If this seems like hair splitting, bear with us. In our experience, nothing produces more confusion than giving someone the grand sounding task of establishing a methodology without being clear about what that entails. Like any project—and the establishment of a methodology is an important subproject in the deployment of a PMO—the most critical step toward success is the first one: defining requirements.

What is required for a company to institute a project management methodology? A methodology standardizes the structure of managing projects, sequences the project phases, and describes best practices so that there is predictability, repeatability (of desired results across the organization), and more efficient utilization of resources. It also reduces the risk of cost and schedule overruns by promoting a deliverable-based program for every project. With a little research, many off-the-shelf methodologies can be readily tailored to your environment and practice, then implemented.[3]

A glossary of project management and process improvement terms compiled by The Boeing Company defines "process" as a systematic series of actions directed to some end ... a series of progressive and interdependent steps by which an end is attained.[4]

One elegantly simple definition of methodology was suggested by IBM's Tony Nish: "When managing projects in a business context, methodology is how we do what we do to ensure high-quality, repeatable results."[5] One of the most immediate values of a standard methodology is that it serves as a common language among practitioners.

Despite this, many organizations lack discipline in the application of methodology. Nish suggests three possible reasons for this:

- They are using a methodology recommended by company edict, but which experienced practitioners haven't bought into.
- They feel it is too high level to be of any practical value.
- They have yet to see the direct positive impact of using one.

How do organizations without a methodology achieve project success? By heroic performance and hiring experienced project managers with a good track record. These "heroes" have long figured out that good methodology is useful and best used when it can be taken for granted. A good methodology is transparent in the hands of an experienced practitioner. It is most simply a noncontent-specific guide for defining and sequencing phases, activities, and tasks. It is, essentially, says Nish, "pure logic that an experienced practitioner applies to meet the needs of a particular business situation or problem." And that's fine for those rare superheroes of the project world, but what about the vast majority of us who are just trying to figure out a better way to manage our projects day in, day out? Imagine an organization with 100 to 1,000 project managers. Do you really want each one to use his or her "pure logic" to come up with a method for managing projects? No. So, let's get into more detail on developing a methodology for the masses.

A methodology should include questions or prompts for all potential issues that might arise in the process of managing a project. This is why methodologies look high level to a nonpractitioner. It is in the *application* of the methodology to the project at hand that the direct practical value emerges.

It is common to hear a project management software package described as a methodology, when it is really simply a tool. The critical difference is that a project management methodology deals with all, or at least most, project management processes in an integrated, orderly way while a software tool offers some project management functionality, but does not cover all processes necessary for success. An uninformed user may not be aware of what it is missing. That's not to say that software cannot play a part; for example, some methodologies are integrated into a project planning tool. Because creating the project plan is a major (and very time-consuming) part of a project manager's responsibilities, using a schedule-creation template that is integrated into a project planning tool and scheduling component can be a useful method, but does not comprise a methodology in and of itself.

"A project management methodology means documentation that incorporates project management processes; teams use a project management methodology as a reference for defining the methods to be used as well as activities to be undertaken during the conduct of a project."[6]

Five Steps to Establishing a Methodology

While this might seem like a daunting task, the good news is that many of the elements of a methodology (Figure 5.1) already exist in your organization. Thus, the first challenges for a PMO when tackling methodology development involve looking around to find out what already is working well and where the gaps are between what you have and what is necessary.

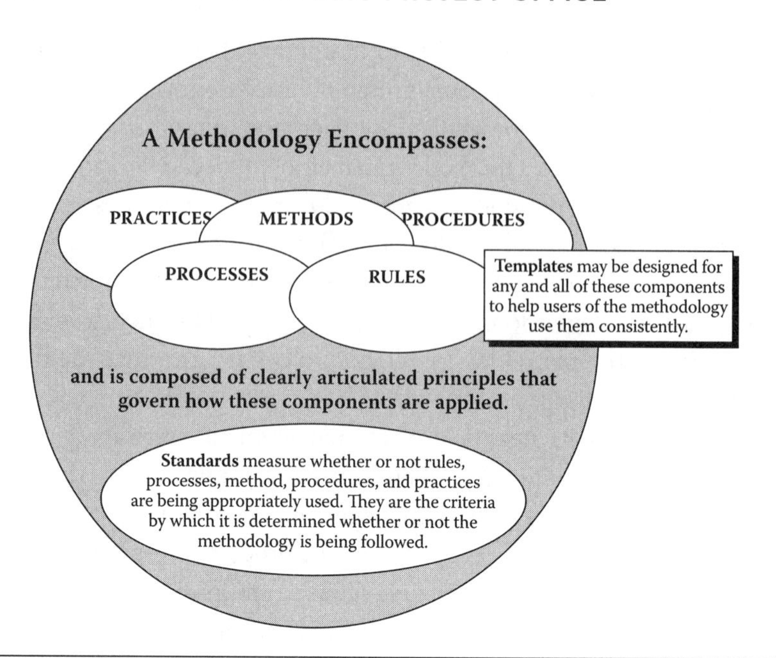

Figure 5.1 What is a methodology?

1. **Map** what's already going on in terms of practices, procedures, methods, and processes. The assessment process described in Chapter 3 offers a road map for this step.
2. **Benchmark** how to manage the types of projects and activities that your company engages in. This is a little tougher and is somewhat outside the scope of this book. While we can discuss best practices in terms of generic project management, which are applicable across industries, there will always be practice standards and methods particular to industry sectors. (See Chapter 10 for a discussion of benchmarking practices.)
3. **Define the processes** that will bring improvements to your company's project management practice. Those processes then can be standardized across all projects by means of rules, templates, and procedures. Establishing measurements (metrics) that reflect whether or not the processes are being properly implemented, and linking those to rewards, plays a major role in the cultural change to a project-based organization. This process audit function is discussed in more detail in the Quality of the Project Management Process section later in this chapter.

4. **Document** all the processes and their component pieces: templates, procedures, and metrics. The resulting compilation of principles, processes, and best-practice tools can properly be called your methodology.
5. **Reassess and refine**. In an iterative process, methodology components will (and should) be in a regular state of review, modification, upgrade, and change. It is a continual evolution as the organization matures in project management practice. A methodology should not be static; if it is, the actual processes that people use in their work will leave the documented processes behind. It is in human nature to tinker with things and try to change and improve them, so no methodology should be allowed to gather dust in a notebook on a shelf. Instead, it should be regarded as a "live" set of documents, describing a lively set of processes. One of the primary functions of the project office is to gather lessons learned and suggestions for improving the project management process from practitioners, and to continuously improve the methodology.

Now you begin to see why the chapter that follows is focused on knowledge management. The collection and dissemination of ideas and information is the backbone of establishing and refining your project management methodology.

What Are the Elements of a Methodology?

Methodology can be likened to a road map that leads project teams from point A to point B during the course of a project. A collection of best practices and repeatable processes, it includes key pieces of information to help project teams succeed. These include:

- An overview of the entire project management process—initiating, planning, executing, monitoring and controlling, and closing—tailored to your unique organizational environment and best practices. This includes project processes, activities and tasks, and the relationships between these elements.
- Checklists of the things that a project team needs to consider, with a description of what is required in each task.

- Key inputs and outputs associated with each of the activities and tasks. These should be accompanied by templates and examples for each of the outputs.
- Guidance on staffing the project team: how many people and with what skills? Templates or decision trees can help identify these players.
- Guidelines for identifying and enlisting the support of the project sponsor.
- An approach for developing project requirements and specifications.
- Standard database structures (a Project Management Information System) for collection, integration, and summary reporting of schedules, costs, and resource usage.
- Guidelines for identification and responsibilities of the project steering committee.
- Guidelines for project portfolio selection, prioritization, and management, including project cancellation procedures.
- Guidelines on application of the project management methodology to projects of various sizes and complexity.

There are two qualities that are characteristic of successful methodologies: (1) they are based on recognized standards and (2) they are flexible and customizable. Every organization is unique and needs to be approached differently due to its internal processes, and different projects will require the methodology to be scalable. When an organization can add its own activities into a methodology and provide guidance to users on how much of the methodology to use for different sizes and types of projects, the ensuing sense of ownership will make it much more of a "live" document, and that will help an organization move to a higher level of capability.

Consistency/Repeatability

In the effort to develop a consistent approach to implementing project management, practitioners have found it easiest to begin with existing internationally recognized standards and models. While by no means the only possible foundations for building a project management methodology, the two that are most commonly referred to, and which we have used in the development of PM Solutions' proprietary maturity model (the Project Management Maturity Model, or PMMM) and methodology center (the Project Management Community of Practice, or PMCOP), are the Software Engineering Institute's (SEI) Capability Maturity Model for Software Development and PMI's *Guide to the Project Management Body of Knowledge* (*PMBOK*®).

The Capability Maturity Model (CMM) In recent years, SEI's Capability Maturity Model (CMM) has evolved to become globally recognized as the best practices standard for methodology development and measurement. Although the principles outlined in the CMM focus on software development process, the CMM approach forms an outstanding framework for establishing a best practices methodology in project management. The SEI says that "for most organizations, the ability to estimate and predict accurately the results of their product development activities from a viewpoint of cost, schedule, and quality is a fundamental business goal. Case studies … suggest that addressing issues of process management, measurement, and institutionalization (i.e., standardization across the enterprise) improve the organization's ability to meet its cost, quality, and schedule goals."[7] How can an organization know how "good" they are at addressing these issues? First, there must be a *standard* against which to measure. That was the concept behind SEI's CMM, which has become the de facto standard model for assessing and evaluating process maturity in the software industry.

On a broader scale, the organizational challenge of estimating and accurately predicting the outcomes of projects of any type or size, in any industry, must address these issues as specifically related to project management:

- **Process Management**. There must be a *repeatable*, quantifiable project management process in place throughout the organization.
- **Measurement**. There must be methods/systems for the measurement of scope, quality, schedule, and cost that are a natural part of running the project, and which are *consistently* applied so that all projects in the portfolio can be compared.
- **Institutionalization (Standardization)**. The processes must be precipitated by the organization's senior management, and *applied uniformly* throughout the organization.

Thus, taking the CMM as our guide, we can collect and document project management processes, assign metrics to determine whether those processes are being correctly carried out, and get management involved in making sure those processes and metrics are uniformly applied throughout the organization. Consequently, we have established a consistent and repeatable methodology for project management.

The PMBOK® Guide The other most commonly referenced document in methodology development is PMI's *PMBOK Guide.*[8] While the SEI research established an intellectual framework for process improvement as related to projects, the *PMBOK Guide* fills in the details by examining what project managers need to know, and what they need to do, to actually manage a project. Many times, companies, having defined a process or set of processes, such as those that are laid out in the project management body of knowledge document, the *PMBOK Guide,* feel they have established a methodology. In fact, according to many experts, they have stopped short of doing so.[9] As the seminal documentation of project management practice, the *PMBOK Guide* is often called upon to serve double duty and be more than it is. There has been some debate among project management practitioners as to whether or not it can properly be termed a methodology. As Tony Nish has written, "While the *PMBOK Guide* describes, at the process level, *what* project management is all about, it does not provide the explicit guidance and tools detailing exactly *how* to manage a project. This confusion between a guide and a methodology," he says, "leads many companies to feel a false sense of security." If you have long-term ambitions for the success of project management as a discipline in the future success of your organization, merely establishing the *PMBOK Guide* as a standards document won't assure your success. You can manage individual projects very well here and there throughout the organization just by knowing the steps and processes laid out in PMI's document. In fact, giving everyone who works on project teams a copy of this document can go a long way toward moving an organization out of the ad hoc state of chaos described in Level 1 of the PMMM. In large measure, this is because it brings a new level of consistency to the vocabulary of people working on projects.

However, a methodology is larger than these foundational elements. A methodology states the standards for processes and knowledge areas, but it also provides, as noted in the section on the CMM above, metrics for judging performance, a system whereby those processes may be consistently applied across an organization, and an overall method—the maturity model—for evaluating the effectiveness of the methodology once it is in place.

A Word about Discipline The other piece of having a consistent and repeatable methodology is, of course, sticking with it. This may prove challenging when you have an organization full of people who are used to doing things their own way. Tying rewards and performance measurement to application of the methodology is one way to gain compliance. For example, on the positive side, a project manager's performance report could state that he or she enthusiastically applied the new methodology and achieved success, while on the negative side, comments in the performance report might show that the individual was slow to adopt the best practices embodied in the methodology, leading to project problems, fragmented team effort, or other indicators.

While rewards and sanctions are an effective way to gain *compliance* after a new methodology is deployed, getting *buy-in* during development is better. Two tips toward gaining buy-in:

1. Get representatives from throughout the organization to act as a steering committee for developing the methodology. Get the most vocal, highest performing representatives you can find. Even critics of the initiative may soon realize their contributions to the methodology are being incorporated and become converts. Roles in methodology development are shown in Table 5.1.

Table 5.1 Roles in Methodology Development

ROLE	RESPONSIBILITIES
PMO director	Oversee PM methodology development Champion PM methodology implementation
PMO staff	Develop content (e.g., templates, forms, etc.) Document methodology
Company project managers	Provide subject matter expertise and guidance Implement PM methodology

Source: Adapted from Jaques, T. Imaginary obstacles: Getting over PMO myths,: www.gantthead.com/articles (accessed March 5, 2001).

2. Second, use the standard methodology as a conflict resolution tool. A formal system that can be invoked when people have complaints or criticisms can help separate out the serious issues from the mere complaints, if it is used consistently. When faced with a complaint, suggest, "Why don't you file a risk statement for that?" or "Submit that to the change control board."[10]

Scalability to Projects of Varying Size and Complexity

If this is sounding very complicated, don't worry. One of the primary concerns most organizations have with establishing a methodology is that, while they know they need a comprehensive approach to managing projects, they also worry about getting into "analysis paralysis." They wonder whether on the many smaller, less complex projects that make up their portfolio, the focus on planning, the project board, the project reviews, risk analysis, change management, extensive reporting, resource allocation, resource control, schedule variance analysis, cost variance analysis, and knowledge management will actually hold up project completion rather than facilitating it.

This is a commonly voiced concern, but the truth is that any good methodology is scalable. It can be used in an extremely rigorous fashion to make large, complex projects a success, but it also should be flexible enough that smaller projects can be carried out under a simpler set of rules. To assist managers of small projects achieve a viable balance between over-managing projects and running unreasonable risks from under-managing them, we developed some simple tools and checklists to aid the project manager in scaling the methodology to the project (Table 5.2).

Table 5.2 How Much Methodology Do I Need?

	SMALL	MEDIUM	LARGE
FIRST TIER EVALUATION			
Effort hours	40–200	200–1000	Over 1000
Elapsed time	Less than 3 months	3– 6 months	Over 6 months
# Project team members	1–6	6–12	Over 12
SECOND TIER EVALUATION			
Technical complexity	Level of integration, skill sets, done before?		
Business complexity	Number of business units, level of business process changes		

Use this table to help determine the size of your project. First evaluate your project per the first tier evaluation. Then think through your project in light of the second tier evaluation. This second tier evaluation may cause you to move your project up (e.g., from small to medium), but almost never down. The numbers can be adjusted up or down, depending on what kind of projects are most common in your organization; however, always keep in mind the finding that small, short-term projects have a far higher rate of success than large or long-term ones.

All project management deliverables (i.e., project charter, statement of work, etc.) need to be produced in some form, even for small projects. The important thing is that some thought has been given to how much of the methodology should be used for smaller projects. The actual amount of information documented for a small project will normally be much less than for a medium or large project. The *minimum requirements* for small projects, whether internal or external, are:

- Clearly defined and justified business needs
- Tightly defined scope and key deliverables
- A documented project plan
- An organized set of resources
- Appropriate control procedures

With small projects, there is a strong temptation to "just do it" without worrying about managing it properly, as called for in the methodology. This approach often leads to project failure due to major midproject scope/deliverable changes and course corrections, or worse yet, production of the wrong, unusable, or unsupportable deliverable.

All projects, regardless of size and scope, need the fundamental processes of project management applied. On smaller projects, risk is lower, the cost impacts are lower, and the impacts of the technology on the delivery of the project may be lower, so the need to engage in rigorous analysis may be lessened. However, risk must still be assessed and costs determined before a decision can be made to move to a more compact set of guidelines.

For example, all projects need to be chartered. All projects should have a project manager assigned, even if part time. All projects need a project plan, including scope, work breakdown structure (WBS), schedule, budget, risk analysis, and control processes put in place. And, a project manager on a small project still should understand the enterprisewide implications of every project, still should be trained in managing projects, and still should have access to a comprehensive methodology. He or she may choose (or be required) to apply only limited segments of the methodology that are relevant to challenges they face on a project; however, in order to choose, the methodology must be available, commonly practiced, and must include a matrix or table categorizing projects and listing the portions of the methodology that must be applied to projects of various sizes and complexity.

Finally, this scalable feature of methodology means that the PMO has a rapid and relatively low-cost way to add value. With a series of templates and a methodological approach to planning and managing projects, a small PMO staff can quickly open shop and begin to show results. Many companies can maximize the expertise and existing technologies to offer a core set of services that can be of great value to the organization. As the organization phases in the full Strategic PMO (SPMO), capabilities increase to match the organizational needs.[11]

Overview of a Sample Methodology

As we noted earlier, PM Solutions' methodology documentation, the *Project Management Community of Practice*, a portion of which we offer here and in the included CD-ROM as an example, is based on experience in the field, on process definitions developed by the Project Management Institute (PMI), and on principles of process improvement laid down by the SEI. We chose these standards documents as foundational material because they incorporate process and methodology standards known to be successful in the initiating, planning, executing, monitoring and controlling, and closing of successful project management initiatives worldwide as well as knowledge from many seasoned project managers.

The Processes

It is a truism that an organization's success is dependent on documenting and understanding the processes in which that organization is involved. It is also axiomatic that the success of any project can be defined by how "satisfied" the stakeholder (or customer) is at the end. In developing a methodology, we set out with the intent to provide all the processes and

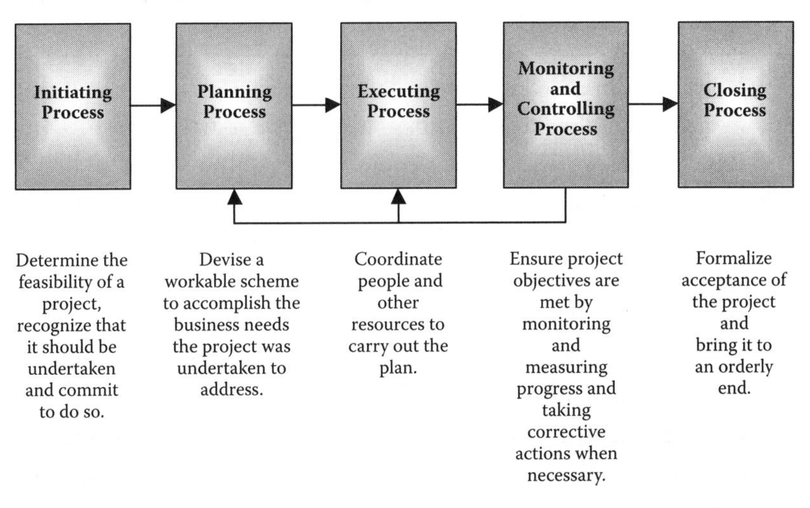

Figure 5.2 Project management processes. (Adapted from *Guide to the Project Management Body of Knowledge*, 4th ed., Project Management Institute, 2008.)

tips necessary for a project manager to manage any project, regardless of size. In order to do that, we must first define these processes; luckily, the PMI has largely done this for us. The overall process of project management can be viewed as five separate but related processes, as shown in Figure 5.2.

Notice that one process feeds the next. Also notice that there is a high degree of interaction between Planning, Executing, and Monitoring and Controlling. Not only are these processes interactive, they also *overlap* as the information (results/deliverables) flows from one to the other; thus, in fact, the interaction isn't as simple as it looks in this graph. The project management processes are not individual, one-time events. Rather, they are overlapping processes that occur at varying effort-levels throughout the project. We won't belabor the descriptions of these processes, as they are laid out in the *PMBOK Guide* in great detail. Because Executing and Monitoring and Controlling are where most of the actual management of the project takes place, let's look at them in detail. The following list provides some factors and guidelines to help you determine how much (and what kind of) ongoing monitoring, control, and adjustment will be required for successfully managing your project.

- **Size and complexity of the project.** Small projects may only require a subset of the control procedures. Large projects will most likely require rigorous control procedures in all areas. If the project has a *large scope* and it is being delivered in a relatively *short period of time,* a lot of project management will be required.
- **Risk of the project**. Risky projects typically require tighter control in those areas of risk that affect the project the most. A thorough risk analysis will identify these areas.
- **Deviation from standards in the project**. Projects with tight tolerances in any one of the "triple constraint" areas—scope and quality/schedule/sesources—require tighter control to ensure that the tolerances are met.
- **Critical nature of the project**. Projects that are delivering functions critical to the business (whether the customer is external or internal) must be controlled tightly, specifically around scope and quality, to ensure that the critical requirements and success criteria are being met.
- **New techniques or technology used in the project**. If the project will be working with new techniques or technologies, there will be more uncertainty and risk associated with the project. This will require more project management specifically in the areas of risk and quality management.
- **Specific end date to the project**. Predetermined deadlines affect how a project is managed and the level of monitoring and control required. If the project has a specific end date that must be met (for business or legal reasons), tight monitoring and control needs to be employed to ensure that the date is met.
- **Regulatory requirements**. Federal, state, and local regulations often require that projects produce specific deliverables and provide evidence that control procedures were in place during the project's life cycle.

Finally, project Closing processes ensure that all necessary project work is completed and that all valuable experience and feedback is captured for future use. A general "rule of thumb" is that project Closing should take no more than 2 percent of the total effort required for the project; however, project Closing should not be minimized, as is often the case, particularly the activities involving lessons learned. There is far too much knowledge and experience that is lost to future projects and project managers by not giving this process proper attention and having a knowledge management "system" to capture the knowledge for future use by all resources (see Chapter 10). Proper capture of lessons learned can be of particular usefulness to the ongoing improvement of your methodology.

The Templates

Tables 5.2 and 5.3 give an overview of some of the templates included in the methodology; this list is scalable and adaptable to your organization's business and size. Not every organization will need every template listed, and not every project within that organization will make use of all of these. A selection of similar templates, derived from PM Solutions' PMCOP tool, has been included with this book as a CD-ROM.

Quality and PM Methodology

Why discuss quality when talking about project management methodology? Because it is such an integral part of project management that it must be covered, in considerable detail, in a project management methodology. After all, organizations implement a methodology primarily as a way to ensure better quality of deliverables. In this chapter, we present an overview of just what we mean when we say "quality." In practice, we find considerable confusion when discussing the topic.

The overall objective of any project is to satisfy the customer; this is the real meaning of *quality*. Yet this apparently simple concept is often misunderstood. Quality in project management has a very specific goal: To ensure that the customer's requirements are being met and that the project team is adding value to the project by adhering closely to the project management process defined in the organization's project management methodology.

The project team is responsible for project quality. By using established project management techniques and exercising an appropriate level of technical oversight, the project manager and project team can

Table 5.3 Typical Templates and Checklists by Process Group

INITIATING	PLANNING	EXECUTING	MONITORING AND CONTROLLING	CLOSEOUT
Business Case Instructions	Project Logistics Checklist	Project Status Report Template	Scope Change Request/ Impact Template	Project Closeout Checklist
Project Determination Checklist	Kickoff Meeting Agenda	Customer Sign-Off Checklist	Change Log	Project Closeout Report Template
Business Case Instructions	Project Logistics Checklist	Project Status Report Template	Scope Change Request/ Impact Template	Project Closeout Checklist
Project Determination Checklist	Kickoff Meeting Agenda	Customer Sign-Off Checklist	Change Log	Project Closeout Report Template
Scope Statement Checklist	Project Risk Log Instructions		Project Control Checklist	Project Sign-off Form
Project Charter Template & Instructions	Issues Log			
Project Binder Contents Checklist	Issues Log Instructions			
	Project Planning Process Checklist			

Note: Templates above, which are representative examples of those normally included in a project management methodology, can be found on the CD-ROM supplement. Many additional templates and checklists are typically tailored and included in an organization's methodology.

make quality a regular, essential part of day-to-day project planning and execution. To assure or control quality, the project team must monitor quality on two dimensions. Let's take a close look at this often-misunderstood concept of quality in both of these dimensions.

To begin, we'll pose two questions. First, if a project manager follows the project management process to the letter (does everything

Table 5.4 Additional Templates and Checklists

INITIATING	PLANNING	EXECUTING	MONITORING AND CONTROLLING	CLOSING
Roles and Responsibilities Checklists: Sponsor, Project Steering Committee, Project Manager, Key Stakeholders	Customer Requirements Checklist	Project Staff Acquisition Procedures	Cost and Schedule Control Procedures	Final Product Evaluation Checklist
Project Manager Assignment Letter	Work Breakdown Structure Template	Supplier, Vendor, and Subcontractor Payment Certification	Vendor/ subcontractor Management Procedures	Post Project Review Template
Project Organization Chart and Guidelines	WBS Dictionary Template		Project Steering Committee Status Report Format	Project Management Process Assessment
Project Requirements Matrix	Cost Estimating Guidelines			
Statement of Work Template/ Instructions	Logic Network (PERT Chart) Template			
Project Initiation Process Checklist	Resource Responsibility Matrix			
	Project Communication Plan Template			
	Project Schedule Review Checklist			
	Project Budget Review Checklist			
	Project Quality Plan Template			
	Customer Relations Plan			
	Test and Acceptance Plan			

exactly as defined by the organization's project management methodology), does that constitute project quality? And, second, looking at it from another angle, if the project manager totally ignores the methodology—doesn't have a kickoff meeting, doesn't build a WBS, doesn't keep others informed of progress—yet produces a deliverable that completely delights the customer, is *that* "project quality?"

Let's face it, the latter event does sometimes happen. What we are interested in when we develop a methodology is to capture the best practices employed by those heroic individuals to whom project management comes naturally. (As an aside, this is also the basis for defining competencies in competency-based management, which we will discuss briefly in Chapter 7). While they may not label their activities the same as we might, chances are good that they are, indeed, adhering to a project management process each time they manage a project. And, to the extent that a formal project management methodology captures the best practices of all those heroic practitioners, the correlation between following the methodology and ultimate project quality, as measured by customer satisfaction, is also high.

Can we measure or assess project quality defined in terms of both the quality of deliverables and adherence to the process? Yes. To measure the quality of deliverables, we perform a "test and evaluation" process of the technical work. To measure the quality of the project management process, we perform quality process audits. We'll discuss each in turn, providing practical guidelines you can use to assess both elements of quality on your project.

Quality of Deliverables: the Product

The goal of any project is to produce a product or service that satisfies the customer's expectations, be it a customer internal to the organization or an external customer. Implied in this statement is the understanding that we actually understand what the customer wants. That is frequently not the case. In fact, meeting the customer's true requirements is not easy, and even satisfying the stated requirement is, in many cases, not sufficient for the following reasons:

- Customers may not completely understand their needs, therefore, they cannot clearly state what those needs are.

- There may be opportunities for additional value to be added through performance of the project, therefore, the requirements may change during the project.
- There may be a disparity between the stated needs and current capabilities.

Obviously to have any chance of meeting the customer's expectations, the supplier and customer must agree on the customer's needs and expectations before launching the project. Because it is so difficult to satisfy a customer's emerging requirements, we should modify our definition of quality to "conformance to mutually agreed-upon customer specifications." The only way to obtain this agreement is through a detailed joint examination of the customer's requirements, a discussion of the supplier's ability to meet those requirements, and documentation of the agreed-upon requirements. Then, to ensure that the project team continues to satisfy the requirement, it must practice rigorous change control throughout the project life cycle. Adequately examining, evaluating, and defining customer requirements is essential to successful project implementation, and it forms the basis for assessing quality as it pertains to project deliverables. How should the project team proceed?

The project team should not force its vision of the requirement on the customer. A better approach is to follow these guidelines:

- Strive to understand the customer's requirements as stated by the customer.
- Determine whether those requirements are achievable in the current environment.
- Identify potential solutions to the customer's requirements.
- Strive to understand the business needs that drive the requirements.
- Identify alternate approaches that may address the customer's business needs.
- Identify its own capabilities that must be expanded to meet the customer's requirements.

Once the project requirements are understood and the customer and project team mutually agree upon and document expectations, project planning, execution, and control may proceed. After the project plan is put together, control of product quality during the execution phase becomes the priority. Monitoring and controlling project implementation is, in many ways, the same as monitoring and controlling project quality.

Earlier, we defined quality as satisfying or conforming to mutually agreed-upon customer specifications or requirements. These specifications are comprised of four main attributes:

1. **Business requirements**. What business purpose the initiative must satisfy and how its value is determined.
2. **Technical requirements**. What the product or service must do, how it must appear, and what performance criteria must be met.
3. **Schedule**. When the project must be completed.
4. **Cost**. What the completed project will cost.

Through customary project control mechanisms, the project manager controls project implementation and, ultimately, its quality. Each of the usual project control mechanisms is also a means of controlling project quality. With an emphasis on controlling the project's schedule and costs, and with continual evaluation of technical progress, the project manager ensures that appropriate work is being accomplished on schedule and within the established budget.

The project manager must establish a routine of evaluating progress and monitoring quality based on reviewing reports from project team members and evaluating the information they contain. For example, using earned value methodology helps the project manager understand quickly whether a project is meeting schedule and cost requirements (two of the four attributes of quality). The project manager assesses the third and fourth attributes, technical and business performance, through ongoing assessment of performance and through a comprehensive testing process.

Quality of the Project Management Process

When an organization creates and deploys a project management methodology, it does so in the belief that those who follow the methodology will achieve better results: on time delivery, within budget, and according to specifications (this was defined as product quality above). This will lead to satisfied customers and the long-term health of the enterprise. Performing project management quality assessments is a way to provide senior managers and project managers with an indication of the health and strength of the project management practices being applied within their organizations.

Project management quality assessments, or audits, can be applied in two ways. The first way is simply to assess whether the organization is following the published, mandatory project management methodology. This is done by developing checklists to determine if the steps in the project management process are being followed as required, and if the project management artifacts being produced (an artifact is a deliverable, such as a project plan, critical path schedule, etc.) are being prepared properly. The audit process requires an auditor, typically from the project office, to conduct interviews with selected project managers and team members, to review the project management artifacts against the guidelines in the methodology, and to score the project against objective criteria contained in the checklists. Audit results are provided to senior management independently for information and corrective action.

Typical project management artifacts include the following:

- **Project charter**. A document issued by senior management that authorizes the project manager to use corporate resources to fulfill the purpose of the project
- **Statement of work**. A document describing the expected outcome of the project, including the deliverables to be produced and what will/will not be included as part of the project.
- **Success criteria**. A document describing the factors by which project success will be measured. These could include schedule and budget performance, return on investments (ROIs) to be achieved when developed product is deployed, and a measure of user satisfaction.
- **Project organization chart**. A pictorial representation of the project team structure. It should include position titles and names of individuals assigned to those positions. It should also identify all user and support personnel.
- **Corporate organization chart**. A pictorial representation of the corporate structure. It should include position titles and names of individuals assigned to those positions. We should

be able to identify the project sponsor and project manager within this organization chart.

- **Responsibility matrix**. A matrix indicating who on the project team is responsible for what. We should be able to readily identify the project's requirements manager, quality control staff, configuration manager, status reporter, schedule manager, etc. Many times these are additional duties assigned to development staff.
- **Job descriptions**. Specific job descriptions for each position on the project team.
- **Work breakdown structure**. A hierarchical representation of all work to be performed as part of the project. We should be able to associate each WBS element to a scheduled activity and to a statement within the project's statement of work. The document should also include the WBS dictionary, with definition of each work element.
- **Product breakdown structure**. A hierarchical representation of the products to be produced by the project. We should be able to associate each PBS element to the WBS elements and to a statement within the project's statement of work.
- **Estimating standards**. The estimating guidelines and actual data used by the project management team to estimate the level of effort needed to complete the project. The standards should identify the type of estimation performed (i.e., lines of code, function point analysis, size of documentation) and any tools used in the estimation process.
- **Project schedule, including baselines**. The documentation that informs all concerned parties when various project activities will begin and end. The schedule should identify who is doing what work and the dependencies within the activities.
- **Weekly project status reports**. Any reports produced for distribution internally within the project team that communicate project status
- **Project plan, including cost and schedule management plans**. The documents that identify how the project will be managed and what the final product developed will be. This document is usually developed during the project planning activities.

- **Project review and approval process**. A description of who will be selected to perform the role of project sponsor and project steering committee members. This documentation will include roles and responsibilities, duties, approval process, and frequency of project steering committee meetings for the project.
- **Deliverables sign-off process**. This is a discussion of how the project will be closed down upon completion of all deliverables or upon cancellation. The document is a management document, not a technical document.
- **Risk identification and mitigation plan**. The document that describes the project risks, their probabilities, and the strategies that will be used to mitigate those risks. It should also describe how risks are identified and quantified on the project.
- **Requirements management plan**. The document that describes how requirements will be defined and managed throughout the project's life cycle. We should be able to identify the project's requirements manager and the requirements control mechanism within the plan.
- **Quality assurance plan**. The document that describes how the organization will evaluate the project's performance to ensure that the project will meet quality standards. The quality assurance manager should be identified.
- **Change control plan (CCB)**. The document that describes how changes to the project's scope will be managed. The makeup of the change control board, including how often it convenes, should be discussed in the plan. Note that the CCB can be the same as the project steering committee.
- **Change requests and log**. The documents used in the project's change control process.
- **Action items tracking documents**. The documents used to track action items for the project.
- **PMO charter**. The document that describes the roles and contributions of the PMO.
- **Subcontractor contracts**. The legally binding document describing the statement of work, including terms and conditions, applicable to subcontractors used on the project.
- **Lessons learned reports**. Any postmortems that have been performed on the project to date.

- **Miscellaneous project documentation.** This umbrella category includes the following artifacts: project status meeting agenda, project meeting minutes, kickoff meeting agenda, and minutes.
- **Visibility (war) room and procedures.** Any documentation supporting the project's "visibility room" if one exists. Today, it usually exists only in cyberspace; this increases, rather than lessens, the need for thorough documentation of procedures for creating, accessing, changing, editing, and storing content.

Each of these artifacts will be identified in the project management methodology, including its purpose, when it should be prepared, instructions for preparing the product, and an example of a completed artifact. The auditor compares what was produced against the standard in the methodology and assigns a rating.

The second form of evaluating project management quality is to use a project management maturity model against which project management practices are compared. Rather than use the organization's project management methodology as the standard, checklists are developed that measure the maturity of the process against the maturity model for each of its sections. Comprehensive interviews are conducted with project managers, team members, customers, other stakeholders, and senior management to assess the overall quality of the project management process and to identify areas in need of improvement. As in the first approach, project management artifacts are examined to complement the information gained in the interviews.

Using PM Solutions' PMMM as a model, a typical assessment report shows scores for each of the nine PMMM knowledge areas, with narrative to support the scores assigned. A project management improvement plan is also a feature of this method. (Appendix C shows an example of such an improvement plan.) Where does the project office fit in with respect to either product or project management process quality? That depends on whether the organization has a separate quality assurance office. If it does, the QA representatives will work with project managers to help them assure the quality of deliverables. When it comes to assessing project management process quality, however, it is more often the project office, which performs the process audits, that reports results to senior management, and works with project managers to improve the quality of their project management efforts. The QA representative advises and assists the project office as necessary.

To briefly summarize the topic of quality, the purpose of developing a methodology is to improve the quality of results that an organization achieves through projects. Process improvement, in a nutshell:

- Defines the gap between stakeholder needs, organizational objectives, and operational outcomes (see the assessment process outlined in Chapter 3).
- Conceptualizes, with input from project knowledge workers, the operational changes that need to take place in how work is done and how value is delivered in order to resolve the defined gaps (see Gap Analysis, Chapter 3).
- Quantifies the value-added that will be realized once the new processes have been implemented. More than ever, this is the key to maintaining a best practice PMO; the ability to track benefits and show how the PMO methodology contributes to business results should be a starting point in the design of any methodology.
- Designs new work processes and procedures that embrace the changes identified.
- Quantifies the time, cost, and plan for implementing the improvements.

These steps will form the planning stage of your PMO deployment project.[12]

Methodology Deployment A project management methodology is essential in portfolio management. Once the methodology is written, it must be deployed and enforced. This implies periodic process audits by the SPMO to ensure it is being used properly. The SPMO itself is a user of the methodology in its role as overall portfolio manager. For all project managers and the portfolio selection team, training must be provided, along with specific guidance on use of the various tools and templates provided in the documentation.

Measurement The SPMO ensures that individual project status is reported and controlled, that individual status is integrated with portfolio objectives in each business unit and across the enterprise, that tradeoffs are made for the benefit of the entire organization, and that relevant information is passed up the chain to senior management. The metrics used to measure performance vary slightly among organizations, but normally address, at a minimum, cost and schedule variance, scope and quality deviations from plans and specifications, and customer satisfaction. (See the discussion of the Project Dashboard in Chapter 8, and of performance measurement in Chapter 10.)

Project Audit This is a critical internal project review, particularly in a complex project. It is at this point that the collective wisdom of the SPMO professional staff can be most helpful in developing the best possible solutions. This internal project review may contain:

- Review of data collection (Are there any gaps?)
- Review of findings and recommendations (Are they supported? Are they complete? Is the logic sound? Can we anticipate client reaction? Have all elements of the proposal been considered?)

How Good is Your Methodology?

In the research project on the quality of project management methodologies proposed by Nish,[13] quality of a methodology can be evaluated on four dimensions:

1. *Breadth* is a measure of how comprehensive the methodology is. That is, how useful it is at all stages of a project and how transferable and generic it is across industries and types of projects. *Test question*: Can the methodology be applied across different project types?
2. *Depth* is a measure of how much detail is provided in each stage or phase of the methodology. Too much depth can be a negative in usefulness for a specific type of project, while too little doesn't provide the practitioner with sufficient explanation to apply the concept. *Test question*: Does the methodology support the development of detailed work plans, controlling cost and managing resources, and all other *PMBOK Guide* elements, for a particular project type?

3. *Clarity* is a measure of how easy the methodology is to use and explain to someone just learning it, either a new project manager or project team member. A high percentage of users of the methodology are satisfied they understand and can explain the methodology if called upon to do so. Words and phrases used in the methodology are readily translatable into words and phrases more commonly used in my industry. *Test question*: Is the methodology understandable and easy to use?
4. *Impact* is the fourth dimension because other criteria don't mean much without results. *Test question*: Does the use of the methodology contribute to better (more timely, cost effective, and/or higher quality) results on projects?[14]

This framework for evaluation is one that can easily be used to take a look at a given methodology once it has been in place for a time, or to assess an off-the-shelf product. As it has not been fully tested, we don't endorse it, but refer the reader to it as a technique of interest.

A good strategy for implementing the methodology is to run a pilot project using it within one functional area of the company, e.g., IT or production. This provides a venue to test and refine the methodology for a given set of project deliverables, while offering an excellent training opportunity for the managers. Managers would do well to choose an area that they know with authority and adopt a well-known methodological approach that can be tailored to meet the specifics of the environment. After a success, the methodology and the PMO staff can move to another functional area of the organization and modify the approach yet again. This course of action complements the low-cost, scalable approach, and is in keeping with the process improvement spirit of methodology development.

Governance: Balancing Order with Creativity

One concern that project managers have sometimes voiced is that, by standardizing processes via rules, procedures, and templates, we are promoting a method at the expense of creativity—standardizing the life out of the "art of project management." This is certainly a concern, particularly in organizations where new product development and other creative endeavors form the bulk of the projects being managed. It should be remembered that the purpose is to improve organizational outcomes. The old saying about rules applies to methodologies here: "It's not knowing when to obey them, it's knowing when to break them." Be willing to bend methodologies when that makes sense for the project or the organization.

Having said that, let me contradict myself somewhat to stress the role of methodology compliance. (After all, it's been said that one sign of a superior intellect is being "able to hold two opposing ideas in mind at the same time and still retain the ability to function."[16]) Adherence to a common methodology is one of the markers of a PMO with appropriate governance in place.

Governance is a concept that is infiltrating the organization from both directions. On the enterprise level, Sarbanes–Oxley regulation has driven the development of corporate governance policies, reengaging top executives with issues of risk, ethics, and finance that had become obscured by layers of bureaucracy. Meanwhile down in the trenches, IT and data governance policies proliferate as technology personnel struggle to keep massive amounts of information organized while doing too many projects with too few people. While the organization without appropriate governance becomes chaotic, it's important not to fall into the trap, as so many PMOs have done over the years, of becoming "the methodology police," and making the role of serving to improve the business into one of enforcing rules. PMOs that are viewed as a helpful resource and asset will be around much longer than those that are hated purveyors of bureaucracy.

Thus, it's important to remember that the central component of governance is *decision making*. Strategic decision making must occur within a policy framework that lays out clearly defined roles, accountabilities, and processes; otherwise, everyone in the enterprise is at the mercy of executive whim. Governance is that framework. It assists the organization's leaders in making the strategic decisions to fulfill the organization's purpose as well as the tactical actions to be taken at the level of operational and project management. In addition, such a framework helps workers and teams to understand the actions they need to take to deploy and execute the organization's strategy. And, it ties everyone in the organization together around consciously chosen purposes. Or, to put it simply, governance is integration—with clout.

Defining Governance

Many people, at least in U.S.-based businesses, are familiar with governance as an IT term: "The assignment of decision rights and the accountability framework to encourage desirable behavior in the use of IT."[17] But, why stop in IT? Shouldn't the entire organization—across all departments and projects—have a policy framework to encourage "desirable behavior?" And, should it be the *same* framework, not one for IT, one for finance, and another for HR? Defining the rules by which the enterprise operates should be done once for the entire enterprise, not piecemeal, department by department.

An initial step in the formation of governance is the absorption of input from the business environment and from stakeholders within as well as external to the organization. This input can come in various forms, for example:

- As process maps of existing processes and/or maturity assessment of those processes
- As existing role descriptions for groups within the organization (descriptions at the level of individual roles are too granular a level for a governance policy)
- From a knowledge base, such as a lessons-learned database
- From focus groups with employees, managers, vendors, customers, or other crucial stakeholders
- From best practices resources, such as benchmarking forums or reports

Management, whether executive or middle, operational or project, provides the link between governance policies and the actual work: the organization of tasks, people, and technology to get the job done.

It's important to remember that "good governance" means not only achieving the desired results, but achieving them in the right way. That "right way" is shaped by the needs, values, and culture of the organization; however, some universal norms do apply. For example, consider these characteristics of good governance from a list published by the United Nations Development Program in 1997:

- **Participation**. Providing all stakeholders with a voice in decision making
- **Transparency**. Built on the free flow of information

- **Responsiveness**. Of organizations and processes to stakeholders
- **Effectiveness and efficiency**. Processes and institutions produce results that meet needs while making the best use of resources
- **Accountability**. Of decision makers to stakeholders
- **Strategic vision**. Leaders have a long-term perspective[18]

Many organizations struggle for a definition of governance that is both comprehensive enough to achieve its purposes, yet simple enough to be easily put into practice. Most experts on governance agree, although they may use differing terms, that there are two major aspects to defining governance: (1) defining processes and (2) defining roles and responsibilities in carrying out the processes. To discuss these two aspects, we will borrow terms used by British IT writer Neville Turbit on his IT governance Web site.[19]

Process Governance

Agreed-upon processes, described at a high level by the organization's governance document, should flow from the top for consistency throughout the organization. Often, when seeking to set forth governance, companies do just the opposite, describing what's already accepted practice in the departments and attempting to roll it up to the enterprise level. In setting forth governance, take it from the top, defining first how strategy is made, and then how that results in the evaluation of ideas, their justification, approval, and prioritization; the commissioning of projects and programs; the roles of the departments in those programs; and of the personnel on those projects. This is the only way to shed light on the famous "gray area."

Methodology and Standards

Process standards, such as the *PMBOK Guide* for project management, may be stipulated in a governance document, but should not be described in detail. Instead, it is common for a governance document to refer to a more detailed document. It may say, for example, that the organization's project management methodology must be followed for projects in any function area that exceed a certain dollar value or resource threshold. Governance development also can drive the development of standards, identifying where a needed standard doesn't exist, and stipulating that this lack should be addressed.

Naturally, governance is also about setting up a system for assuring compliance with the policies. There's no point in having rules unless there's some process to ensure that people are following them. This can be formal, as in a governance audit, or informal, as in periodic reviews. The important thing is that the right metrics are collected so that performance and compliance can be monitored. The purpose here is not to punish, but to tweak processes, develop personnel, and change strategic course if needed. Participating in a well-governed organization ought to be a rewarding experience.

Which brings us to the second area that must be addressed in governance.

People and Structure Governance

In order to execute strategies effectively, people need to understand their roles in making strategy happen. A strategically focused culture is one in which the organization structure, and the defined roles of groups and individuals, are designed specifically to smooth the way toward goal accomplishment. Many organizations, due to inchoate strategies or the buildup of obsolete roles or functions over time, actually have elements in the management and work layers of the structure that hinder and impede progress towards strategic goals instead. When putting a new governance structure in place, it's a good time to clean house—to examine the organization's structure, processes, and roles for outmoded and unhelpful aspects.

A Governance Framework

In our book, *Seven Steps to Strategy Execution* (Center for Business Practices, 2007), we defined a set of responsibilities and practices exercised by the board and executive management with the following goals:

- Strategic direction is clearly understood throughout the organization and business units and levels of management are all focused on aligning to this strategic direction.

- Strategic objectives are achieved. This is controlled by establishing a continuous loop process for measuring strategy performance, comparing to objectives, and redirecting activities or changing objectives where necessary.
- Appropriate and effective processes are in place to monitor risk and that a system of internal control is effective in reducing those risks to an acceptable level.
- Verification that the enterprise's resources are used effectively and efficiently.
- Decision makers have the information necessary for making decisions.[20]

A governance framework means that all strategic decisions throughout the organization are made in the same manner. Each level within the organization must apply the same principles of setting objectives, providing and getting direction, and providing and evaluating performance measures. A common governance framework ensures that decisions are made the same way up and down the organization and that there is an appropriate mix of people making decisions. Longtime practitioners of project management will instantly recognize in this a path around the "portfolio decisions based on popularity" paradigm that has for so long plagued our organizations.

Best Practices in Governance

Our 2005 research study, *Strategy and Projects*, identified six best practices related to governance that had been proposed in management research. They are:

- The organization has a well-defined strategy.
- A documented strategy execution plan guides strategy execution efforts.
- Strategy is communicated clearly to those developing portfolio and program/project plans to ensure that those initiatives support the organization's strategy.
- Portfolio, program, and project managers feel a sense of ownership about the organization's strategy execution plans.
- Appropriate and effective processes are in place to monitor and manage risk.

- Decision makers have the information they need about the execution of their organization's strategy to make optimum decisions.

Not surprisingly, the research results indicated that high-performing organizations use governance best practices more than other organizations, consistently and significantly, while low-performing organizations consistently underutilized governance best practices. It's worth noting, in this chapter on methodology, that high-performing organizations are significantly better than average at having appropriate and effective processes in place to monitor and manage risk.[21]

Notes

1. The purpose of this chapter is not so much to present our own methodology, although we will draw our examples and templates from it, but to discuss methodology, processes, and standards as general topics, and how the PMO supports and offers these. We are striving to be more descriptive than prescriptive.
2. *The American Heritage Dictionary of the English Language*, 3rd ed. (Boston, MA: Houghton Mifflin Company, 1992).
3. Tim Jaques, Imaginary Obstacles: Getting Over PMO Myths: www.gantthead.com/articles (accessed March 5, 2001).
4. Ibid.
5. Tony Nish and Jeannette Cabanis, "The Consulting Methodology Survey," PMI. *PM Network* (August 1999).
6. Ibid.
7. Software Engineering Institute. *Capability Maturity Model, Systems Engineering Improvement* (Pittsburgh, PA: Carnegie Mellon University, 1995): 2–27.
8. PMI Standards Committee, *A Guide to the Project Management Body of Knowledge*, 4th ed. (Newtown Square, PA: Project Management Institute, 2008).
9. Nish and Cabanis, "The Consulting Methodology Survey."
10. John Sullivan, "Hidden Roles of a PSO," *PM Network* (February 2000).
11. Jaques, Imaginary Obstacles.
12. Michael Wood, "What Is a Process Improvement Methodology Anyway?": www.Gantthead.com/articles (accessed June 2000).
13. Nish, "The Consulting Methodology Survey."
14. Ibid.
15. Jaques, Imaginary Obstacles.
16. F. Scott Fitzgerald, quoted in *The Oppposable Mind: How Successful Leaders Win through Integrative Thinking* (Cambridge, MA: Harvard Business School Press, December 2007).

17. J. Kent Crawford, et al., *Seven Steps to Strategy Execution* (Glen Mills, PA: PM Solutions' Center for Business Practices, 2007).
18. Institute for Governance materials: http://www.iog.ca/boardgovernance/html/gov_wha.html (accessed June 2007).
19. Neville Turbit, "IT Governance Project Governance," http://www.projectperfect.com.au/info_governance.php (accessed June 2007).
20. Crawford, *Seven Steps to Strategy Execution.*
21. Center for Business Practices, *Strategy & Projects: A Benchmark of Current Best Practices* (Glen Mills, PA: PM Solutions' CBP, 2005).
22. Jeannette Cabanis-Brewin, "Project Failures: Making News, Driving Change," *Best Practices Report Executive Briefing* (Summer 2000).

6

Project Portfolio Management and the Strategic PMO

In the first edition of this book, we felt it necessary to start by defining Project Portfolio Management (PPM). Today, PPM is ubiquitous in project management circles, even if it is sometimes poorly understood or considered to be a software function. A few years ago, it wasn't uncommon to read or overhear arguments about whether or not there was a difference between managing multiple projects and PPM. That conversation seems to have come to a natural end, as both project management practitioners and executive leadership begin to realize that, while managing projects well (both individually and collectively) is critical to the organization, managing the project portfolio is another set of skills altogether. Think of it this way: If each project is an instrument in your organizational orchestra, you have programs made up of the strings, percussion, and wind instruments, but someone—the conductor—has to optimize the whole system, and turn it into something greater than the sum of its parts. It may be worthwhile to take a quick look at how PPM has evolved before we explore the critical role played by the Strategic PMO in implementing and "conducting" PPM.

The Evolution from Project to Portfolio

Project management's focus on planning and controlling the activities of a project met the needs of organizations in the past because most identified projects were large-scale, long-duration, single projects—a skyscraper or a legacy computer system. With the shift to knowledge-based wealth creation, today's projects have a shorter, market-driven life cycle and more intangible deliverables. Companies have projects of all sizes, projects that involve all areas of the organization in some capacity and often involve people outside the organization as well, in the form of outsourcing suppliers, contract labor, consulting firms, or public stakeholders. It isn't surprising that a disconnect has developed between companies' ability to manage projects on the project level, and their ability to strategically align them on the organizational level.

Project portfolio management provides a consistent way to evaluate, select, prioritize, budget for, and plan for the "right" projects, those that offer the greatest value and contribution to the strategic interests of the organization.[1] Doing the right thing starts with developing a strategic focus and ends with project selection; that's PPM. Doing the right things *right* (and quickly) is project management.

However, you can't manage what you can't measure, the old saying goes, and unless all the projects on the table can be held up to the light and compared to each other, a company has no way of managing them strategically, no way of making intelligent resource allocation decisions, no way of knowing what to delete and what to add. Project portfolio management brings order to this situation in two important ways:

1. It brings realism to an organization's planning processes by aligning what an organization *wants* to do with the resources—the money, hours, people, time, and equipment—required to *get it done.*
2. It brings rationality in the allocation of resources, both human and financial.

Project managers have taken a lot of heat when project management didn't deliver organizational nirvana, but the truth is that the business of selecting which project to invest in must be carried out at the executive level, at the level of *managing by projects*, rather than of *project management.*

As project management knowledge and practice has matured over the years, organizations have gotten more skilled at the management of projects, and in a natural evolution, their focus has begun to shift, from intraproject planning and execution to the front-end of the project management life cycle: how to choose the correct projects, how to prioritize them, how to track the prioritized projects, and how to ensure that there are adequate resources to staff those priority projects. This front-end work is PPM, and it is as necessary to projects as projects are to the portfolio. Even projects that succeed in meeting their requirements can be failures if the project requirements don't match up to the business realities faced by the company. For project management to live up to its promise, projects must be selected with care.

While most decision making on the project level is concerned with tactical issues (How can we do this thing right?), decision making on the portfolio level is concerned with strategic issues (How can we be sure we are doing the right things?). Today's emphasis on PPM is part of a general trend toward systems thinking in organizational life. Instead of tweaking the parts (individual projects, departments, or processes), systems theory encourages us to look first at the whole: the enterprise.

A frequent refrain in the business press is that projects must develop more of a business focus. The same might be said of business; a project focus is required. Without a project focus at the highest level of the business, projects seem to pop up at will across the organization, generating confusion. There's a lack of clarity as to how projects align and link to organizational strategy; often there's no business process for selecting projects, and senior management is unaware of the number, scope, or benefits of the projects being undertaken. As a result, people both on projects and on the business side feel they are working at cross purposes with each other.

Giving projects a strategic focus goes a long way toward resolving these concerns. Combining a strategic focus with a business process for selecting and prioritizing projects is an important step in creating an environment for successful projects.

The move to PPM, therefore, is a necessary step up the ladder of organizational maturity. The place of portfolio management in a mature organization is depicted in Figure 6.1.

No one would think of building by turning 100 different construction teams loose to build 100 different rooms, with no single blueprint or agreed-upon vision of the completed structure to go by. Yet, this is precisely the situation in which many large companies find themselves with regard to managing by projects. Companies routinely over-schedule their resources (human and otherwise), run redundant projects, and damage profitability by investing in off-topic activities that don't contribute to organizational health.

Applying effective PPM practices is becoming increasingly important to all business entities. Small or large, each organization must

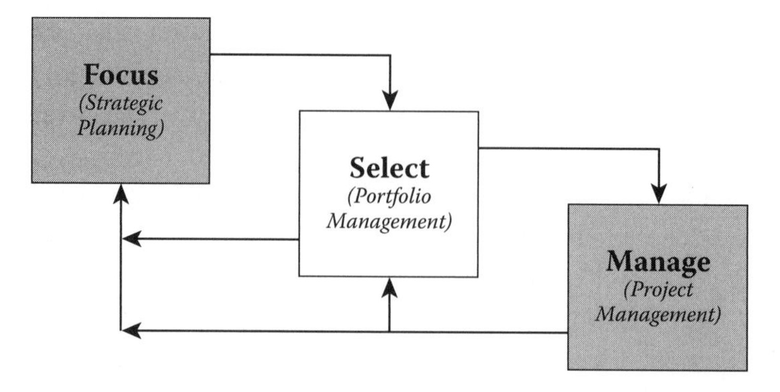

Figure 6.1 Selecting the right portfolio is only one element of success. First, the organization needs to focus, via strategic planning and strategy management. The strategic focus of the organization becomes the foundation for selecting research, new product development, information technology, and business improvement projects. Second, the right projects need to be selected. Third, after selection, the projects/programs need to be managed well.

select and manage its investments wisely to reap the maximum level of benefits from their investment decisions. In the information technology (IT) investment arena, IT executives are finding that senior management is demanding a tangible return from their large investment in information technology.

Organizations in many other industries are implementing project management practices to improve their ability to execute projects on time and within budget. Just as important as improving project management practices and organization, however, is the need to execute the right projects. Implementing an effective PPM process helps organizations select those projects, supporting business goals, diversifying the investment, and maximizing business performance.

The Business Case

On the negative side of the case, companies without good portfolio management are prone to a welter of problems (Table 6.1).

The lack of defined process for project selection often leads to the handshake method of approving pet projects outside the budget cycle. These projects generally come down from executives and, in the absence of an approval process, project managers find it hard to refuse. In addition, there are many maintenance and regulatory projects that operate in the background; they must be done, but have not been consciously budgeted either from a time or money standpoint.

Table 6.1 Common Problems in Portfolio Management

- Too many projects "off strategy"; disconnects between spending on projects and strategic priorities
- Too many unfit, weak, mediocre projects
- Poor projects are not killed, but take on a life of their own
- Resources are scarce, but unfocused
- Too many trivial projects (updates, modifications) and not enough major breakthroughs

Source: Adapted from Cooper, R.G., Edgett, S.J., and Kleinschmidt, E.J. 1998. *Portfolio Management for New Products*, Reading, MA: Perseus Books.

Naturally, where the difficulty shows up is in the frayed nerves of over-scheduled resources. The lack of a formal process for selecting — or killing — projects is a primary driver of project manager burnout. As the project teams become more frustrated and overworked, even less is accomplished, so that low-priority (or no-priority) projects drain the resources needed to complete high-value work.

In fact, the high-value, strategic or breakthrough projects that will serve as the company's future revenue generators often cannot even be started because scarce, unfocused resources become overscheduled with low-value tasks. In summary, the formal, mindful selection of projects has benefits to all levels of the organization, from the strategic to the administrative.

On the positive side, PPM provides a solution to many of the problems that commonly plague projects and the companies that depend on them.

Project Failure Applying PPM to projects helps to resolve some of the key issues that lead to project failure. Of the top ten factors leading to project failure, five (incomplete requirements, lack of user involvement, lack of resources, unrealistic expectations, and lack of executive support) are addressed by the implementation of a system that engages corporate leadership in a structured process of selecting and prioritizing projects.

One immediate benefit of PPM is the ability is to choose and prioritize projects in such a way that responsible decisions can be made as to which projects to kill, and when. Jim Johnson, chairman of the Standish Group, has identified portfolio management as *the* process that can make the difference in project success. Said Johnson, "Companies need a process for taking a regular look at their portfolio of projects and deciding, again and again, if the investment is going to pay off. As it stands now, for most companies, projects can take on a life of their own."[2]

Johnson stressed that continuous self-assessment should be built in, allowing earlier kill decisions on failing projects, with the associated cost savings. This frees up money and personnel for dedication to projects that are worth pursuing. Killing a project, said Johnson, should be seen as "successful resource management, not as an admission of failure."

Financial Performance Proper portfolio management results in bottom-line yields; another reason why it's important for the executive level to champion this process. And project portfolio management should be an easy sell. Research by Cooper, Edgett, and Klienschmidt[3] reveals that, for R&D portfolios, the top 20 percent of companies had an explicit, established method of portfolio management, consistently applied across the organization. The same research showed that even rudimentary portfolio management processes created a spike of benefits almost immediately. Of those in the top 20 percent, about 70 percent had used the portfolio management method for more than two years. So, even businesses that are relatively new to portfolio management begin to see positive results very quickly.

Time to Market Toney and Powers found that in large functional organizations that implement a project management group and its corollary practices (including portfolio management), lead times to market have been reduced by as much as 60 percent, development costs have declined, quality has improved, and forecasting accuracy has increased.[4]

Risk Management In addition, assessing and managing risk becomes easier within the context of a project portfolio, according to research by E. L. Jarrett.[5] To deal with any significant risk, there must be diversification, along with other kinds of balance to ensure continuity and health for the enterprise, such as investing in lower-risk projects that provide a near-certain return, or investing in higher-risk projects, the path to extraordinary returns.

Apolitical Decision Making Jim Johnson has observed, "All failures are political."[6] When projects are selected and planned in an ad hoc, chaotic environment, objectivity about value goes out the window. Johnson stressed that having a standard methodology helps to take the politics out of decision making. "Instead of opinion, we have process."

A *Research Technology Management* article by Jarrett listed the factors that can delay—or conceal—the need to terminate projects. First on the list was "personal pride."[7] PPM methodology cleans up the project selection process by offering a checklist of criteria for project approval.

"Chunking" for Success Big projects, simply, are more prone to failure. Jim Johnson credited the rising project success rate to "very small projects that get going very quickly ... you get to the important milestones very quickly ... if you are failing you can tell quickly if you need to kill it. You can act to save time, money, and resources." Susan Cramm, in a *CIO Magazine* article, described the process necessary to reorganize around small projects: Reframing big initiatives as a series of smaller $1 million to $2 million initiatives, but warned that "this cannot happen in the absence of a PPM methodology."[8]

Achieving Strategic Goals Ideally, an organization would conceive of, fund, plan, and monitor its projects through a "strategic lens." Without such a holistic view, it's easy to have lots of activity going on without much of it being of real value. (Value isn't necessarily profit, but can include the development of intellectual capital, the improvement of public image, or movement toward a long-term evolution into a new kind of business.) Mistakes, innovations, even losses can all come into focus through this lens. For project-oriented companies, that lens is PPM.

Better Asset Management PPM is simply managing your assets. In a way the project management language/angle on portfolio management is a distraction. Portfolio management has less to do with project management skills and more to do with strategic planning. Of course, in order to make sound decisions on a portfolio level, decision makers must have sound information on the project and program level.

Better Resource Management Even if your company hasn't been downsizing in search of short-term profits, top-performing subject matter experts aren't exactly a dime a dozen. How can you make sure your smartest people are working on your top-priority activities? This is a two-sided issue: one side impacts corporate profitability, the other impacts employee morale (which, by the way, impacts corporate profitability, according to RHI's (Kilgore, TX) Human Capital Index).[9]

In short, PPM is the action of a mature organization to gather all the pieces of the project value puzzle—human resources, technology, corporate strategy, and financial resources—into a coherent picture. That's why enterprise resource allocation is a key feature of PPM. It shows whether projects can be staffed with current resources. If the resources are not available or if they have been committed to more work than they can reasonably accomplish, projects will not be completed as scheduled. This may sound simplistic, but in practice, many companies do not know how many projects they have scheduled, or who is going to do them. PPM sheds light into this darkness.

Overall Benefits

The benefits of portfolio management, which apply across the board, to government, not-for-profit, and for-profit entities, include:

- A structure to select the right projects and remove the wrong projects.
- A system for placing resources where it matters and reducing wasteful spending.
- A method for linking portfolio decisions to strategic direction and business goals.
- Establishing logic, reasoning, and a sense of fairness behind portfolio decisions.
- Establishing ownership amongst the staff by involvement at the right levels.
- Avenues for individuals to identify opportunities and obtain support.
- A way for project teams to understand the value of their contributions.

And, according to research sources compiled by Kendall, it can offer:

- 20 to 30 percent improvement in time to market
- 25 to 300 percent improvement in number of projects completed with the same resources
- Average project duration cut by 25 to 50 percent
- Over 90 percent project success rate, with double the profit margin
- 50 percent improvement in R&D productivity[10]

PPM in the Strategic PMO: Where Strategic Planning Meets Project Execution

Although PPM has long languished within IT departments among "those project management types," there is simply no way for top-level corporate leadership to hand off leadership on the project portfolio management issue. That's because choosing the groups of activities that create value is inherently a strategic activity. And, the basic issues that arise when developing a portfolio management system (not a software system, but a collection of processes—a PPM methodology) are issues that touch on organizational development. Jim and David Matheson identified some of the issues as:

- How can we aggregate our opportunities into manageable strategic projects?
- Who is the overall process owner? Who will facilitate the analytical process?
- How do we get top management and project leadership to buy in to the results?
- How will business and marketing units interact in the process? Should we use one cross-functional decision team for the entire portfolio, divide responsibilities by business or technical areas, or use a multilevel review structure?[11]

These are not "project management" questions. The management of the corporate portfolio of projects is a strategic management issue. In summary, project portfolio management provides a consistent way to evaluate, prioritize, select, budget for, and plan the right projects;

those that offer the greatest value and contribution to the strategic interests of the organization. When used effectively, portfolio management ensures that projects are aligned with corporate strategies and priorities that the portfolio contains the right mix of projects, and that resource allocation is optimized. It is the practice that bridges the gap between the executive decision process and project execution.

The Strategic PMO: An Organizational Home for PPM

The SPMO is the optimum organizational solution for aligning projects to strategy and tracking project and portfolios to ensure they continue to meet the needs of the business, even as these needs continue to change over time. It serves as the critical link between business strategy and execution of tactical plans.

In high-performing organizations, the SPMO owns the PPM process.[12] The SPMO ensures that an organization's projects are linked to strategic plans. It may be involved in facilitating the prioritization and project selection processes, and typically is intimately involved in resource allocation decisions. The SPMO also coordinates tracking of the current portfolio, analyzes portfolio performance, and is instrumental in administering the stage-gate process for all projects.

Portfolio management begins with the selection of the portfolio. Just as it is beyond the scope and charter of individual project managers to manage a portfolio of projects, it is beyond the scope of a Type 2 Project Office to manage a corporatewide portfolio. That leads to the need for a Strategic PMO and is, in fact, one of the primary reasons for the existence of the SPMO. It is the "voice of the projects" on the executive-level steering committee that must decide which of the many opportunities to pursue with a limited amount of resources. The decision of which projects to authorize is complex and depends on a number of factors, such as return on investment, fit with the current portfolio, desire to introduce a new product line, availability of resources, and many others. This is a classic executive decision, involving many decision criteria and alternatives. The first task for the committee is to select a decision support tool to help organize and simplify the decision process. This can be done via a proprietary tool or with commercially available software decision tools. The importance is not the method, but the organizational will to organize around strategy execution.

The SPMO is the optimum organizational solution for aligning projects to strategy and tracking project and portfolios to ensure they continue to meet the needs of the business, even as these needs continue to change over time. It serves as the critical link between business strategy and execution of tactical plans.

The literature on project portfolio management, while vast compared to what was out there even three years ago, is still new enough to be confusing and, in many cases, full of untried optimism and untested theories. Writers stumble over each other to offer the three elements, four phases, five steps, and so on. From the executive leadership point of view, all these checklists miss the point. The question that must be addressed when an organization commits to PPM is not merely which tools to buy or which mental model to use. It's a question of maturity. Only organizations mature enough in project management practice to implement a PMO can succeed at portfolio management and only mature Strategic PMOs fully realize the strategy execution promise of PPM.[13]

As shown in Figure 6.2, the place in the organization where strategy and PPM meet most seamlessly is in a Strategic PMO. The

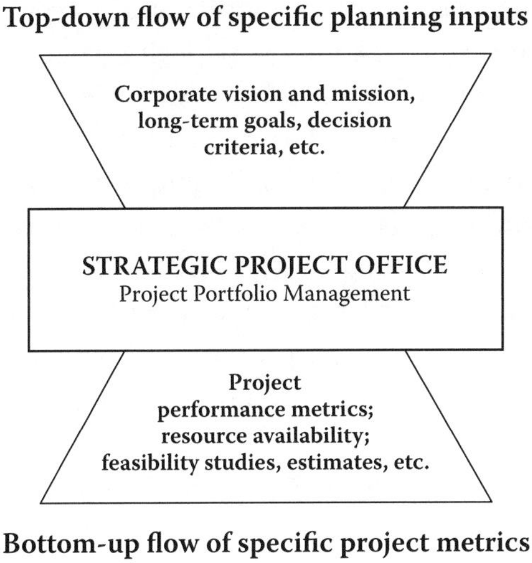

Figure 6.2 The Strategic PMO integrates strategic planning and project execution through project portfolio management.

development of an enterprise-level project office and the implementation of portfolio management go hand in hand. If an enterprise-level project management office does not own the process of project inventory, prioritization, and selection, it cannot be done well. The META Group (now part of Gartner, Inc.) recommends this strategy, and those companies that have put enterprisewide PPM in place, such as Cabelas and Northwestern Mutual Life, have relied on it. In fact, while intradepartmental portfolios may perhaps be selected and balanced without involvement of a project management office, it's doubtful that anything on a wider scale can succeed, and you can't optimize the system by balancing only parts of it. This is undoubtedly why Gartner predicted years ago[14] that companies failing to establish a project office would experience twice as many major project delays, overruns, and cancellations as will companies with an SPO in place.

Just as project management can fail without senior leadership involvement, PPM can fail for lack of project manager input. Without the input and buy-in of project personnel, it is difficult to maintain continuity between the selection and execution phases. An enterprise-level PMO, with personnel dedicated to portfolio tracking and management, answers this lack.

When the commitment is made to PPM, your company can effectively execute on projects that are meaningful to your organization's vision, increasing your chances of achieving your long-term goals while executing on important short-term projects. Without PPM, the long-term results can get lost in concern for short-term results. Whenever corporate leadership shifts from a long-term common vision and stretches goals to a short-term horizon, the organization suffers. Short-term goals must be balanced with vision, values, plans for market leadership, and management commitment to long-term growth. The absence of long-term goals and vision are particularly unhealthy for the firm's future: breakthrough projects, R&D, and the competitive horizon.[15]

A noted writer on strategic execution, George Veth, calls PPM "initiative management," and this is perhaps an effective way to think of it when discussing the important role that PPM plays in execution. The key components of initiative management that tie into strategy management include:

1. Explicit funding of the investment in strategic initiatives.
2. The adoption of a formal and rigorous project management methodology.
3. A periodic review of the measurable impact of initiatives on the business.

PPM also can be defined as a decision-making process, in which a set of alternatives is evaluated against defined priorities and critical tradeoffs are intentionally considered and determined. This definition highlights three important aspects of the PPM: generating a pipeline of ideas, establishing priority criteria, and making tradeoff decisions.

Governance and PPM

The term *governance* is often used in the context of PPM to refer to the ongoing process of decision making associated with keeping the portfolio healthy and with ensuring that the processes set up for identification, validation, and selection of projects are followed. This "portfolio governance" exists within the larger framework of organizational governance.

At the level of organizational governance, once the PPM processes that will be used organizationwide have been defined, an executive dashboard should be sufficient to keep work on track between portfolio decision meetings. It's important that key metrics associated with PPM be identified in order to make the dashboard as useful as possible while incorporating as few metrics as necessarily. One "death knell" for PPM lies in swamping C-level executives with the project-level minutiae.

To carry out the role of liaison between executives and the portfolio, the role of portfolio manager has become more and more accepted. The role is discussed in more detail in Chapter 7.

Once roles have been defined for PPM, and the selection process codified, the primary role of governance in PPM is to promote the uniform application of the methodology across the organization. Finally, the approach used to analyze projects must be uniformly applied across the organization. The criteria and weights should be identified, defined, and completely documented. The process established should ensure that project teams consistently interpret the criteria and its related measures and/or probabilities. An analogy can be made to performance reviews and the use of criteria and performance scales. For example, different managers may have different definitions of an "excellent" rating. In order to ensure fairness for all employees, it is imperative the organization clearly defines all performance levels so managers consistently interpret and apply the scales. If this doesn't happen, employees may be unfairly evaluated and rewarded, just as projects may be unfairly evaluated and rewarded. If the cost, benefits, and risks are identified for one project, they should be identified for all.

Assessing Organizational Readiness

Is your company mature enough to do portfolio management? An organization that is has to have certain basic organizational attributes and infrastructure in place. The organization should:

- **Have a coherent strategy.** Portfolio management uses project activities to move the organization forward toward a goal. "Groups and individuals that conduct strategic planning," say Frank Toney and Ray Powers, "are consistently more successful at achieving goals than those who don't utilize it. Research also indicates that maintaining a constant focus on the goal of the group has a high correlation with goal attainment."[16] Have a clear, shared vision.
- **Know how to manage projects.** Walk before you run. Trying to implement portfolio management before mastering the basics of managing individual projects is putting the cart before the horse. Without project management methodology and practices in place, you don't have the most basic data with which to work. Bad project management means cost and schedule estimates that are exercises in fantasy.
- **Know what projects it has.** What's in the portfolio? Many companies don't know. A complete list of all the initiatives competing for resources is a baseline requirement to even begin portfolio management. Many companies set criteria for what counts as a project to be listed (e.g., only projects that surpass a predetermined threshold number: a schedule of 30+ days or 100+ hours, or a budget of at least $50,000). Counting

projects is a first step toward deriving value from portfolio management because certain realities are quickly revealed. If you schedule 130 percent of your human resources to projects, for example, a lot of things will not get done, and valuable people will quit or go crazy.

The good news is that even this most basic step surfaces redundancies and dead issues, allowing a portfolio management initiative to create value for the company almost immediately. The inventory has to include *all* projects because resources are working on all projects, not just on the high-profile ones. And, it should include projects that are being carried out by outsource providers and consultants, as well, because even those projects have at least someone within the company as a liaison, contract manager, or project manager. These hours often get "lost" in the decision-making process, only to show up later as a cost or schedule overrun.

- **Be fully described.** Once the parameters of the list are decided upon, each project on it must be described. Details, such as technologies required; estimates of time, cost, and personnel required; and a basic risk/reward calculation, give portfolio managers the data they need to compare and contrast projects. Companies skilled at opportunity identification on one hand and at tracking existing projects on the other, have a significant advantage at this stage. Their lists will be more complete and their estimates more meaningful. Many questions must be answered in detail before you can begin to select and prioritize the projects in the inventory, and some of the answers won't come easy. Which projects make the most money? Which have the lowest risk? Which have subjective value, in terms of community image or internal morale? Which are not optional—projects dictated by regulatory requirements, for example? In the information-gathering process, a second level of shakedown will naturally come about. Some projects will be held off because human resources aren't available, some because the technology is immature, some due to looming external risks. Much, if not all, of the information necessary to populate the database about projects in the inventory is generated by the steps in the project management process: scope definition, risk management, scheduling. The software and resources required to generate, analyze, and deliver this information should not be scattered all over the organization, but unified under the flag of the Strategic Project Management Office.

- **Have reliable project data.** You won't make good decisions based on bad information. This is where the enterprise-level project management tools with portfolio management capabilities really earn their keep. Gathering data in such a way that it can be put into context and become information is where software reigns supreme. How long will each project take? How much will it cost? What's the expected return on investment (ROI)? What's the status on the projects already underway? Being able to view the most salient information on each and every project in thumbnail sketch form allows executives to compare apples to apples, and weigh the relative benefits of apples versus oranges. What is less certain is organizational willingness to use these features of the software, and train appropriately for them.
- **Know who is available to work on projects, and when.** A second part of the inventory process should be, literally, counting heads. Who are the project managers? Surprisingly, many large companies are only beginning to get a handle on who their project resources are and where they reside on the organizational chart. For some companies, the scarcest resource isn't money but project managers. A critical factor in project selection thus becomes: Do we have a project manager who can manage it?

  How many project managers and project team members do you have? What is each one doing, right now? When will he or she be finished with it? What are his or her areas of particular expertise? In fact, without a system for knowing what each person in the pool of potential project personnel is capable of, and when they will be available, you cannot really be said to manage a portfolio. People do projects. Without them, all you have are ideas.

Table 6.2 Have You Asked These Questions?

The following "quiz" will quickly reveal how well your company is managing the project portfolio.

- Does your portfolio reflect and support your business's strategy?
- Is each project consistent with business strategy?
- Does the breakdown of your project spending reflect your strategic priorities?
- Is the economic value of your total portfolio higher than what you've spent on it?
- Once projects start, is there very little chance they'll ever be killed?
- Are projects being done in a time-efficient manner?
- Are your success rates and profit performance results consistent with expectations?
- Is you project portfolio heavily weighted to low-value, trivial, small projects?
- How are opinions of senior people and key decision makers in your business captured in order to make project decisions?
- Have you considered what the right balance of projects for your new product portfolio is?
- Are there redundant projects being performed?
- Have all the projects in play been justified on solid business criteria?
- And of those that were approved, are they still justified?
- Do the managers and team members know where the projects they are working on fit into the priority ranking that best supports the business?
- Are there enough resources to get the work done, and, if there are not, what trade-offs need to be made?
- Which projects make the most money?
- Which have the lowest risk?
- Which have subjective value, in terms of community image or internal morale?
- Which are not optional: projects dictated by regulatory requirements, for example?

Adapted from Amir Hartman, *Portfolio Knowledge,* Aug. 2002.

And, of course, by now it should be clear that, only under the auspices of a Strategic PMO can all the above features be brought into sync.

Table 6.2 offers a quick self-assessment quiz to point up the strengths and weaknesses of your organization's present PPM practices.

The Fundamental Components of PPM

Within PPM are fundamental practices that include linking strategy to project prioritization and selection, and balancing an organization's portfolio to achieve the best results. Also included, typically, are stage/gate models, a process for defining and estimating new projects, communication and review processes for the portfolio, project tracking and reporting, and continual portfolio realignment. Organizations need to apply these fundamental processes and develop a standardized approach to create and manage a diversified portfolio of projects. The stakeholders for the PPM process include financial management, senior business executives, and ultimately the stockholders of the organization as well as employees, vendors and customers. It is a holistic view of all the work taking place and work planned for in an organization.

PPM consists of a number of fundamental components that must be integrated to successfully implement and institutionalize the process. These components can be grouped in various ways. Following are a few ways to conceptualize these. While somewhat overlapping, they all address different aspects of this complex set of processes.

Six Key Processes

Portfolio management is accomplished through the application and integration of portfolio management processes, such as:

- **Inventory**. A process for capturing project data and organizing for portfolio analysis.
- **Analysis**. A process for aligning projects to business strategy, examining business and project risks, then selecting and prioritizing projects in the portfolio.
- **Planning**. A process for approving and funding the project business plans, allocating resources, and scheduling projects. Funding and resource allocation must be based on the identified priority of the project.
- **Execution**. A process for executing the portfolio of programs and projects by means of budgeted resource allocations; focus on getting the work done efficiently and effectively. In order to execute quickly, in the correct sequence, the organization must adhere to the project capacity. Any organization that is overloaded with too many projects sees a dramatic increase in resource multitasking or sharing with a devastating slowdown in project flow. Quick execution also demands that PPM effectively monitor project execution to ensure that out-of-control situations are speedily recognized and acted upon.

- **Monitoring/control**. A process for tracking a portfolio as programs/projects are executed, detecting problems or changes in underlying premises, and reporting to appropriate management levels.
- **Portfolio improvement**. A process for making necessary adjustments to the portfolio, not once but iteratively and formally so that rebalancing and analyzing the portfolio becomes simply the way business is done.

Four Areas of Integration

Four areas that must be integrated in PPM include: (1) budgeting and forecasting, (2) prioritization, (3) resource allocation, and (4) portfolio review and communications. Under the SPMO model, the PPM process is also integrated with strategic planning processes as well as yearly business planning and budgeting processes. Figure 6.3 provides an example of integration between business and PPM processes.

Budgeting and Forecasting An important step in creating and executing a PPM process is to identify all current projects or programs,

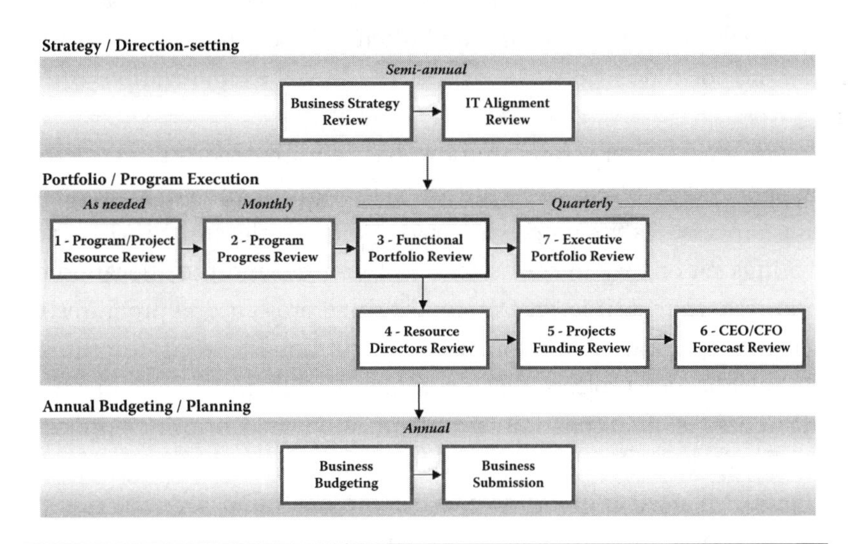

Figure 6.3 Integrated PPM and Business Processes. The timing of the reviews shown vary from organization to organization, but one key to good PPM is that the project mix is adjusted as necessary in a timely fashion.

establish budgets (dollars and human resources), and define start and end dates for these initiatives.

The first step is to *collect* all initiative or project ideas. For organizations already up and running with PPM, this constitutes a regularly refreshed pipeline of ideas for business innovation and improvement. For those new to the process, it is often necessary to inventory and collect all initiatives currently running in the organization, along with proposed projects, so they can be truly prioritized from a strategic perspective. Which projects can be considered strategic? Generally, they are projects that support innovation, organizational improvement, or mandated corporate or regulatory priorities. As Gerald Kendall notes, "If the list is greater than 50 projects, this is a red flag that the organization is not focused." Collect any project plans associated with these projects, including resources allocated, budgets, earned value information, and document who the sponsor is. Determine the justifications for each project and get a status on each active project (this should be high level, e.g., red, yellow, or green).[17]

Future projects or programs also are forecasted and added to the organization's potential portfolio of work. Through years of observation, we have noted that this process is best implemented through careful design of a process that includes the business sponsors, managers providing the staffing for the projects, subject matter experts associated with each project or program, and often a liaison group that ensures that the portfolio meets the business needs of the organization.

Prioritization Every organization has resource constraints, even though leaders often lose sight of that truth in the strategic planning process. As we noted earlier, one of the values of PPM is that it brings rationality to the selection of projects and the allocation of resources. In fact, it is *essential* to prioritize projects and programs to help in the allocation of resources. Prioritization is crucial to optimize the selection of projects that maximize benefit to the organization within its resource constraints. Organizations may prioritize projects based on financial criteria, strategic criteria, or a combination of both. Typical financial criteria include net present value, internal rate of return, economic value added, etc. Strategic criteria may include factors such as alignment with business strategy, customer need and satisfaction, competitive advantage, etc. Often, advice on PPM gives the impression that portfolios are selected, then prioritized. In fact, this is somewhat of a concurrent and iterative process. Inventorying projects will provide some immediate opportunities to eliminate duplication and dead issues. Prioritizing what's left still normally leaves the organization with a "select" but untenable number of active and proposed projects, however. Then, selection can begin in earnest. Priority plays a major role in which projects make their way into the portfolio.

Priorities represent a filtering system for evaluating initiatives. Mapping initiatives against strategies is the first step in identifying viable candidates. Establishing strategic fit can save significant resources (financial *and* organizational) as projects that are nonstrategic are dropped.

Ranking the strategic initiatives according to their business value requires analysis of the business case. (George Veth's equation for this is "benefits less costs, adjusted for timing.") This step is also the basis for assessing the impact of initiatives on budgets and financial forecasts.

The final selection of projects represents an intentional decision to commit resources to fund or staff projects and should reflect the best use of resources based on management's agreed-upon criteria. All this sounds very orderly. But, as one researcher has noted, there are three common problems with the way projects are sanctioned in most organizations:

1. Goals set by senior executives are *not measurably tied* to projects. For example, a vice president claims that a project is essential to meet a goal, but the *percentage* of the goal that the project will accomplish is often not identified.
2. The collection of active projects is *not formally tracked* to see if it is meeting goals (on time and magnitude of improvement promised). Many projects, even in major companies, lack valid, resource-based project plans.
3. Organizations breed *many projects that are not sanctioned* by any executive far too often ... perhaps in as many as 70 percent of organizations.[18]

To address these problems, Kendall recommends the following set of considerations for portfolio balancing:

- **A focus on market and customer needs (as opposed to internal improvements).** If the company has cash flow or other serious financial issues, internal improvements might be the desired "imbalance." However, an organization cannot cost-cut itself to long-term health. It must grow its business.
- **Short-term versus long-term gains.** When too many projects spend money this fiscal year without bringing benefits until the next fiscal year, or sometime in the future, this is a red flag. A portfolio manager should be asking tough questions about benefits realization.
- **Research versus development.** To have a secure future, every organization must invest some of their project resources into research, such as market research, experimentation with new methods, tools and processes, training and human development, motivation, and other areas.
- **Focusing project dollars and human resources on strategic assets.** Assets are not just bricks and mortar. They may include a Web site, customers, external sales agents, distribution channels, and so on. Determine the distribution of project investments to the organization's top five strategic assets, and determine whether or not the distribution makes sense.
- **Balanced sponsorship (IT vs. other functional areas).** In some organizations, 70 percent or more of projects in the portfolio may be sponsored by IT. This is a red flag indicating a lack of balanced ownership of project initiatives.

The use of financial criteria is recommended in most academic discourse on the subject because calculation of these financial factors involves considering all strategic factors in estimating the project profit potential. In practice, however, estimating profit from projects is often difficult, especially for projects that involve a new concept, market area, technology, etc. For these instances, the use of a carefully crafted decision model that is used to prioritize projects based on important strategic criteria is more effective. Two such PPM models, though by no means the only approaches, are described below.

The Fit, Utility, and Balance Model Portfolio selection and prioritization is concerned with fit, utility, and balance (Figure 6.4).

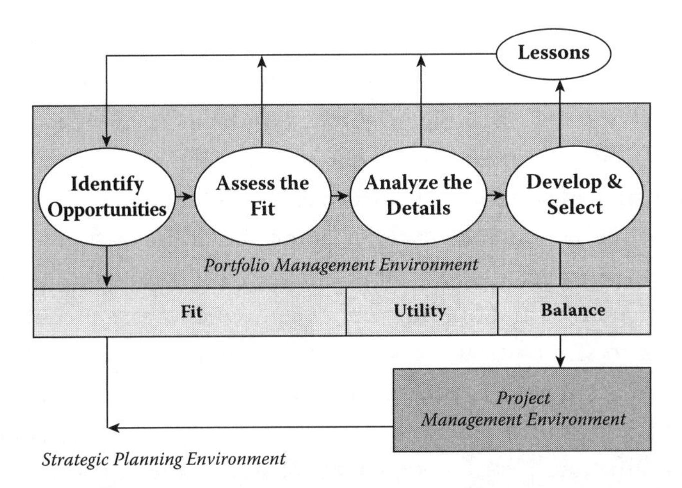

Figure 6.4 How portfolio planning steps merge with the fit, utility, and balance model. Almost every organization will flow through this thought process to build a portfolio of projects. The organization may first identify opportunities, then assess the organizational fit; analyze the costs, benefits, and risks; and finally select a portfolio. The methods and techniques employed differ. Invariably, at some level of sophistication, all organizations will understand the fit and utility of their projects and make some attempt to establish a balanced mix of projects.

Fit—The first major element of portfolio management is to identify opportunities and determine if these opportunities are in line with the corporate strategic direction. This may be the identification and initial screening of projects before more in-depth analysis is conducted. What is the project? Does the project fit within the focus of your organization and the business strategy and goals? Some recommended actions associated with determining Fit would be:

- First, make sure clear strategic direction and business goals have been established because, in an "execution environment, strategic focus becomes the foundation for selecting projects."
- Develop a process to identify opportunities and make it simple and easy to use all of the time. Many individuals have an aversion to complex processes and bureaucratic paperwork. Establish an avenue for communicating ideas versus a paperwork nightmare. Identify a team to review the opportunities and assess the fit within the strategic direction and business goals.
- As part of the process, establish a template for project justification. The ideas need to have some substance and content, otherwise it will be difficult to screen the projects. The template

may include things such as a description of the project, the sponsor, the link to strategy and business goals, and a high-level description of the project's costs, benefit, and risks.

- Establish minimal acceptance criteria. The individuals submitting ideas and the review team should understand the minimal acceptance criteria. There should be basic requirements a project must meet before they are considered for further analysis and funding. Such requirements may include the link to strategy, business threshold minimums (e.g., return-on-investment or cost/benefit ratio minimums), compliance with organizational constraints (e.g., existing technology architecture), and completion of the project justification template.
- Reward ideas and suggestions. Most employees love recognition for their ideas and suggestions. Take the time and interest to formally acknowledge ideas that meet the "fit" test. Always give credit where credit is due.

Utility—The second major element of portfolio management is to further define the project (if needed) and analyze the details surrounding its utility. The utility of a project captures its usefulness, its value, and is typically defined by costs, benefits, and associated risks. Why should this project be pursued? What is the usefulness and value of the project? Several things to consider include establishing criteria to support decision making. Multiple projects vie for resources and funding, and somehow a decision has to be made on which ones to select. To help in the decision-making process, establish common decision criteria and measure each project against the criteria. Because most decisions are based on multiple factors, weigh each criterion to establish its relative importance. This will identify what is most significant to the organization, and allows you to calculate a score for each project using its value for each criterion and applying the relative weights.

Balance—The third major area of portfolio management is the selection of the project portfolio. Which projects should be selected? How does the project relate to the entire portfolio, and how can the project mix be optimized?

- Establish a process that will help optimize the portfolio, not just the individual projects. Industry approaches for developing portfolios range from simple ranking based on individual project financial returns to more complex methodologies that take into account the interrelationships between projects. Regardless of the chosen method, the objective should be to optimize the portfolio, not necessarily the individual projects. The appropriate method depends on an organization's strategic direction, guiding principles, capabilities, limitations, and complexities.
- When selecting the portfolio, consider all types of projects: research, new product development, information technology, business improvement, etc. Remember that relative comparisons are being made, not specific comparisons.

Using a Portfolio Scorecard Model In order to make relative comparisons between projects, some intellectual framework is required. The Balanced Scorecard measures the health of the organization by looking at what's already in place. Doesn't it make sense to use a Portfolio Scorecard (Figure 6.5) as a framework for selecting what *will* be done?

Resource Availability People, money, technology, equipment ... ?	**Profitability** Short-term good or **Necessity** Regulatory or "light-on"
Social Value Internal: Employees, shareholders External: Community, customers, partners	**Strategic Value** Long-term good Support of "horizon" goals

Figure 6.5 A scorecard approach to portfolio selection and balancing.

A Scorecard for Selection might include (each item corresponding to a quadrant in the scorecard):

1. **Resource availability.** Do we have the people, the funds, and the technology to carry out this project? If not, is it worth it to us to develop them?
2. **Profitability.** What are the short-term benefits, usually measured in financial metrics?
3. **Social Value.** What will this project add in terms of image, morale, intellectual capital, and "social equity." Are there intangible benefits internally, for employees and shareholders, or externally, for customers and the community at large?
4. **Strategic Value.** What long-term good do we hope to gain from this project? How does it feed into the vision of the company?

And, in each quadrant of the scorecard, another two questions should be asked: What are the risks associated with this project (also measured in terms of resources, profit, social value, and strategy)? And, what is the learning value of the project in each area?

Viewing projects by means of this scorecard, a company simply can make relative judgments. A project with high short-term profits that carries a high risk of alienating personnel, and not much learning value, may find itself outranked by a project with lesser short-term profitability, but great learning value, as an investment in the future.

There are many other decision methods and tools to help organizations compare a project against the decision criteria, develop a relative score for comparison against other projects, and build a portfolio.[19]

Resource Allocation Resources include capital, expenses, staff, time, or anything that is consumed in the planning and execution of a project. Once an organization prioritizes its projects, it must decide how to assign these limited resources to the projects. Numerous approaches are in use today. One approach arranges projects in descending order of priority and funds projects starting with the highest priority moving down the list until all resources are consumed. Essentially, this technique can be visualized as drawing a cut-line under the final project where resources are fully expended. This approach is often described as the "cut-line approach."

Another method used for allocating resources strives to optimize the benefits to the organization. It involves allocating resources to the group of projects that yield the highest total benefit within the current resource constraints. Specialized techniques can be used in this approach, such as linear programming and cost–benefit analysis. This method is often described as the "optimization approach."

Another important consideration in resource allocation is the importance of allocating resources across a diversified portfolio. Modern portfolio theory, first described by Harry Markowitz, economist at the University of Chicago, advocates investment in a diversified portfolio to reduce risks and maximize expected return.[20] Additionally, businesses invest in a variety of different types of projects, which are often very difficult to compare with each other with respect to business benefits. For example, comparing investments on infrastructure or projects that are targeted at "keeping the business running" are difficult to compare with strategic projects that may provide new markets to the organization. Therefore, an important consideration in the resource allocation process is to define and use different investment categories to help align the total portfolio of investment with the needs of the enterprise.

Portfolio Review and Communications No process, technique, or tool alone is capable of selecting the best portfolio of projects for an enterprise. Processes, techniques, and tools are used in conjunction with senior management's insight, knowledge, and decision-making ability to select the right portfolio of projects. The process, techniques, and tools help to formulate the problem and facilitate the analysis of alternative solutions. Senior management uses this information to select the appropriate projects and to communicate the portfolio to their organization. Because review and communication is so important to the process, organizations should carefully plan project reviews, program reviews, portfolio reviews, and resource planning activities. These reviews provide a forum for studying the alternatives and help to build organizational buy-in for the selection. To establish standards for this process, a number of organizations have developed enterprise calendars that are used for scheduling the reviews at monthly, quarterly, or yearly intervals.

Establish portfolio decision meetings to make decisions. After analyzing the project information and validating it, through application of the decision criteria, the ball should be punted to senior leadership.

Separate decision meetings and teams should be established to make portfolio decisions, using the validated project information. These meetings and reviews are normally held in conjunction with the corporate planning schedule.

Even after selecting the route, it is easy to get distracted along the way; things may not unfold as originally planned, a competitor introduces a new product, the chief scientist makes a key breakthrough elsewhere, legislation is enacted causing industrywide upheaval, or technology advances and matures. Your original selection is no longer the best. So, an adjustment is made and a better path for the future is selected. This is the art of PPM: doing the right thing, selecting the right mix of projects, and adjusting as time evolves and circumstances unfold.

How Do Organizations Implement PPM Practices?

You don't have to wait until all the above criteria are satisfied, so long as there is an enterprisewide recognition of the need to establish these conditions. Any improvement a company makes in managing the portfolio—even if it is as simple as taking inventory—will yield immediate value. Here are some suggestions from the experts:

Start Small and Keep It Simple

Introduce new processes in a phased approach that address the most important activities, including estimation, selection, prioritization, developing performance metrics, and defining a decision process. Smith and Reinertsen's research on the "fuzzy front end" indicated that even otherwise well-managed companies can fail abysmally at opportunity management due to the complexity of their process for dealing with new ideas. In one example cited in their book, the fuzzy front end at a startup company was 500 times faster than it was at a Fortune 500 company, where bureaucratic planning and budgeting processes "guaranteed defeat."[21]

Typically, organizations need to progress through three distinct stages to establish and conduct PPM, as shown in Figure 6.6.

Stage One: Preprocess In the preprocess stage, organizations develop processes used to understand and articulate business strategy and to define, screen, and prioritize projects. They define project characteristics

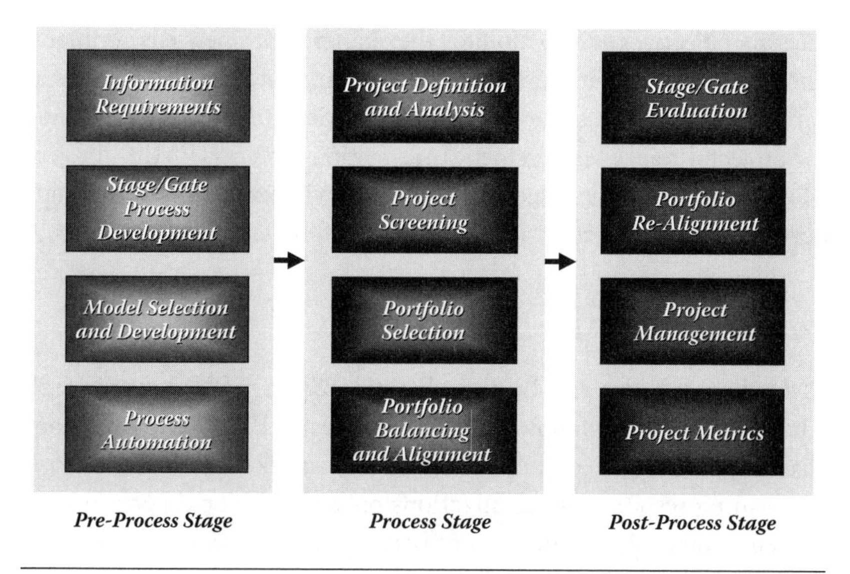

Figure 6.6 Designing and implementing PPM processes.

and informational needs, develop prioritization models and criteria (e.g., financial, strategic, etc.), and establish stage/gate processes (go/no go decision points) for guiding the selection and initiation of projects. This stage involves input from executives, operations/business management, and project-oriented staff. During this stage, facilitation techniques, workshops, and participatory design are used to develop the decision models and stage/gate processes associated with PPM. These decision and stage/gate processes must be designed to align with an organization's culture, functions, and structure, and are best deployed using sound change management techniques. These techniques help to establish management buy-in and to institutionalize the processes.

Stage Two: Pilot and Refine In the second stage, PPM processes are piloted and refined to meet the needs of the organization. The initial PPM process is typically conducted as part of the yearly planning exercises, or aligned with quarterly review cycles. Analysis tools and techniques and portfolio review sessions are further refined in this stage to support management's decision-making processes.

Stage Three: Iterative Evaluation The last postprocess stage includes applying stage/gate evaluation to iteratively assess how well projects are meeting objectives and to monitor the changing business environment. The stage/gate evaluation may be used to terminate or delay projects at various milestones, rescope the efforts, or articulate support to continue the project. Project Management Information Systems (PMIS), Project Management practices, and management review sessions directly support stage/gate evaluation and portfolio realignment in this stage.

Best Practices for PPM

PM Solutions' Center for Business Practices *Strategy & Projects* research study identified the following best practices in PPM. Organizations that consistently used these practices also ranked among the top-performing respondent organizations on organizational performance as well as on project management performance metrics.

- A list of current projects (active, proposed, on hold) is documented.
- Projects are prioritized using a scoring system that uses strategic alignment as a criterion to determine the priority of the project with respect to other projects.
- Metrics are captured to assess the performance of the project portfolio.
- Performance results of the project portfolio are communicated to stakeholders.
- Reviews of portfolio performance and changes in the business environment may cause decision makers to realign the portfolio (killing projects or putting them on hold, reallocating resources).
- Enough resources are in place to make the project portfolio achievable.

The most often used practice by high-performing organizations is having a documented list of the organization's current projects. They are also significantly better than average at having enough resources in place to make the project portfolio achievable. Enough resources! I can hear my readers exclaiming from here. This is the perennial challenge on projects. Yet, the Strategic PMO has some answers to the resource management challenge as well, which we will cover in the next chapter.

Notes

1. Lowell D. Dye and James S. Pennypacker, *Project Portfolio Management: Selecting and Prioritizing Projects for Competitive Advantage* (Glen Mills, PA: PM Solutions' Center for Business Practices, 1999).
2. Jim Johnson, "Turning CHAOS into SUCCESS," *Software* (December 1999).
3. Robert G. Cooper, Scott J. Edgett, and Elko J. Kleinschmidt, *Portfolio Management for New Products* (Reading, MA: Perseus Books, 1998).
4. Frank Toney, *The Superior Project Organization* (New York: Marcel Dekker/CBP, 2001).
5. E. L. Jarrett, "The Role of Risk in Business Decision Making, or How to Stop Worrying and Love the Bombs," *Research Technology Management* (November 2000).
6. Johnson, "Turning CHAOS into SUCCESS."
7. Jarrett, "The Role of Risk ..."
8. Susan Cramm, "Organizational Physics," *CIO* (August 2002).
9. Watson Wyatt, Inc. *The Human Capital Index*: http://www.watsonwyatt.com/research/resrender.asp?id=w-488&page=1 (accessed February 2010).
10. Gerald Kendall, "Project Portfolio Management," in *The AMA Handbook of Project Management*, 3rd ed. (Chanute, KS: AMACOM, 2010).
11. Jim Matheson and David Matheson, *The Smart Organization* (Cambridge, MA: Harvard Business School Press, 1998).
12. James S. Pennypacker, *The State of the PMO, 2007–2008* (Glen Mills, PA: PM Solutions' Center for Business Practices, 2008).
13. Ibid.
14. M. Light and T. Berg, "The Project Office: Teams, Processes and Tools," *Gartner Strategic Analysis Report* (August 1, 2000).
15. Gary L. Tritle et. al., "Resolving Uncertainty in R&D Portfolios," *Research Technology Management* (November 2000).
16. Toney, *The Superior Project Organization.*
17. Kendall, "Project Portfolio Management."
18. Ibid.
19. Meredith and Mantel's Project Management: A Managerial Approach and Pennypacker and Dye's Project Portfolio Management are two resources that discuss the different types of models, and provide a variety of mathematical techniques and examples of how they are used in organizations.
20. Harry Markowitz portfolio theory is summarized at: http://www.thecsem.org/content/basics-markowitz-portfolio-theory.
21. Preston Smith and Donald Reinertsen, *Developing Products in Half the Time* (New York: Van Nostrand Reinhold, 1991).

7

THE STRATEGIC *PEOPLE* MANAGEMENT OFFICE

Human Capital and the PMO

In our first edition, the chapter on roles and responsibilities focused strictly on the individuals involved with kicking off the PMO implementation project. We mentioned some potential roles involving project manager competencies and career paths in a later chapter, but, at the time, we realized much of this was still in the category of "blue sky" for most organizations. At the time, only a tiny percentage of PMOs managed their own personnel. However, in the ensuing decade, the Strategic PMO has morphed into an entity, as described in the Introduction, that manages project managers, schedulers, controllers, and planners as well as projects, programs and portfolios. Therefore, a good deal of new information has surfaced around the SPMO's role in people management that could fill a book of its own—and it does.[1] We will not attempt to cover in this chapter all the important issues involved in managing and retaining project personnel. Instead, we will provide a high-level sketch of just some of the new responsibilities taken on by the SPMO for the "care and feeding" of project managers and other project resources.

The Resource Crunch

Perhaps the most commonly heard complaint from PMO directors and staff is that there simply are not enough trained, qualified people to go around. In 2008, we explored this problem in our study, *Resource Management Challenges: A Benchmark of Current Best Practices*.[2] Even fully staffed PMOs struggle with resource management; the majority of organizations in the study were at a low level of process maturity in resource management. Some of the top challenges noted by study participants included:

- Resource capacity planning is poor.
- Effort estimation is inaccurate.
- Resource risks are not assessed.
- Not enough appropriately skilled resources.
- Resource use is not optimized.
- Schedules/deadlines are unrealistic.
- Resources are assigned inconsistently.
- Too many unplanned requests for resources.
- Contingency planning is inadequate.
- Resource utilization is poorly documented.
- Shifting resources to respond to problems.
- Transition process for resources is inadequate.[3]

Obviously organizations still have a long way to go in learning how to manage project resources for optimal organizational performance. The SPMO plays a role in this by matching scarce resources to the right projects via project portfolio management, and by growing the project talent the organization requires via a well thought out professional development program.

People Make the PMO

It's an old project management axiom that "people do projects." In the old matrix-style organization, those people were fragmented and isolated, scattered across the organization and bedeviled by conflicting priorities. In 2006, our *Project Management: The State of the Industry* research indicated that 46 percent of project offices were overseeing or managing project managers, so we expected the "State of the PMO" research to show the role of the project manager shifting dramatically. And, indeed, in 2007, 60 percent of the responding organizations reported PMOs that manage a staff of project managers. The study also showed a sharp uptick in the numbers of supporting roles that report within the PMO, including planners, controllers, and business relationship managers. In addition, it revealed some correspondences between organizational performance and the structuring of project management roles. High-performing companies were more often implementing enterprise PMOs, and pulling into the PMO a number of roles and responsibilities related to human resource management. For instance, the top performers in the study, far more frequently than low-performing companies, perform resource identification; develop processes for assigning resources; manage a staff of project planners and controllers *within* the PMO; and do their own training, professional development, and performance evaluations. Among the Top Ten PMO functions, according to the study, coaching and mentoring project managers ranked third, with 81 percent of the responding PMOs performing this function.[4]

As of 2007, the average PMO had eight people reporting to it, though the largest PMOs in the study had staffs in the hundreds. And, significantly, PMOs in high-performing organizations have slightly larger staffs than those in low-performing organizations.[5]

We believe this is because the centralization of project management talent, training, and execution under the auspices of a Strategic PMO realizes benefits related to streamlining, eliminating duplication, and allowing project personnel to focus on their primary areas of expertise. It simplifies the tasks of allocating and leveling resources across the portfolio, and introduces realism to the process of hiring, training, and rewarding project personnel.

In Chapter 2, we mentioned several roles of the SPMO identified by Gartner, Inc. Here's why those roles include important implications for "people management," not only within the project office, but across the enterprise:

As developer, documenter, and repository of a standard methodology, the SPMO provides a common language and set of practices that helps to boost productivity and individual capability and takes a great deal of the frustration out of project work. Our research has revealed that over 68 percent of companies that implemented basic methodology experienced increased productivity, and 37 percent reported improvement in employee satisfaction. One manager noted, "If the organization values project management, that translates into very specific policies and behaviors—policies and behaviors that affect the lives of individuals. What is it that keeps people working for a company? It's being valued."[6]

As a center for the collection of data about project human resources (HR) and tools for evaluating and scheduling them, based on experience from previous projects, the SPMO can validate business assumptions about projects as to people, costs, and time; it is also a source of information on cross-functional project resource conflicts or synergies. The human capital implications of rational allocation of human effort are immense. Having common corporate data on resource projections means that planning can be accomplished in a common database, resource projections summarized at the project level, then at the organizational level, on up to the corporate level. This resource-based approach to planning is an integral part of portfolio management, and a key to the rational allocation of resources, both human and financial.

As a center for the development of expertise, the SPMO makes possible a systematic, integrated professional development path and ties training to real project needs as well as rewarding project teams in ways that reflect and reinforce success on projects. This is quite different from the reward and training systems presently in place in most organizations, which tend to focus on functional areas and ignore project work in evaluation, training, and rewards.

As a "competency center" for project management, the SPMO provides a knowledge management locus not only for project management knowledge, but for knowledge about the content of the organization's projects. With a "library" of business cases, plans, budgets, schedules, reports, lessons learned, and histories, as well as a formal and informal network of people who have worked on a variety of projects, the SPMO is a knowledge management center that maximizes and creates new intellectual capital. Knowledge is best created and transferred in a social network or community, and the SPMO provides just that. Through mentoring both within the SPMO among project managers, and across the enterprise to people in all specialty areas, knowledge transfer about how to get things done on deadline and within budget is facilitated.

Through new channels of communication set up by the SPMO, it becomes possible for the entire organizational culture, from chief executives all the way through project teams, to communicate in a common language and work together to understand the issues surrounding how projects are faring and how the issues on one project affect other projects and, ultimately, the organization.

A study of successful project-focused organizations by Oxford University's Christopher Sauer found that organizations with strong project management capability display some key organizational features related to human capital management:

1. **Project-centered role design**. Project managers are assigned sole responsibility for the conduct of the project and, at the same time, sufficient authority is delegated to them to deliver on that responsibility. Given the resources and authority to match their project responsibility, project managers are held accountable by their superiors for the extent to which they achieve targets. But, at the same time, a "project director role" is created and positioned as the immediate superior of a number of project managers. This role encompasses total responsibility and accountability for all phases of the company's involvement with a particular client's project, from business development through to postimplementation.
2. **Valuing project managers**. Successful project-centered companies treat their project managers as an asset, retaining them in bad economic times, even when there are no projects to be managed. These companies are more tolerant of mistakes. With the expectation of continuity of employment and a sense that the company values the project manager, greater openness, innovation, and learning results in more accurate estimating and reporting, and in the preservation of organizational knowledge. Methods, procedures, and standards structure the project management task and formalize some organizational knowledge, but are not expected to substitute for project managers' active skills, which often incorporate much tacit knowledge.
3. **HR management policies that are project-friendly**. Project managers are graduated through a sequence of projects of increasing complexity and difficulty with appropriate culling.

> Since much of this apprenticeship or mentored learning takes place "just in time," it happens not under the purview of HR-based trainers, but within the community of project managers. Consistent with the development of a talented body of project managers, the valuing of project managers, and the tolerance of understandable mistakes, the performance review practice also is more flexible, and project managers have significant input, compared to functional managers, in staff appraisal.[7]

Thus, aligning strategy with training, managing, measuring, rewarding, and promoting people is critical to effective strategy execution. The "people management" issues that best-practice PMOs engage in fall under five interrelated categories, all of which the reader can find discussed in detail in our book, *Optimizing Human Capital with a Strategic PMO* (Auerbach, 2005):

1. Staffing: Candidate identification/selection and/or outsourcing
2. Competency identification and assessment
3. Training (program development and administration)
4. Performance measurement and rewards
5. Career paths/leadership development

Let's take a brief look at each of these before we tackle the team dynamics of the project of rolling out the SPMO.

Staffing the Strategic PMO

How the Strategic PMO is staffed is determined in large part by the role your organization expects it to play. If the PMO is to play the central role in guiding project management in the organization (a Level 3 Strategic PMO), staffing is complex. The PMO director's position within the organization should be equivalent to that of a high-level functional manager, even a vice president, in some cases. The SPMO director at this level is supported by large numbers of professional associates and administrative personnel. One of the perennial issues in execution has been the way the roles on the project level have been defined. The traditional "art/science" clumping of skills and abilities in the project manager has often proved to be nearly impossible to successfully pull off. Refining role descriptions and providing technical and administrative support to project and program managers sets the organization up for success.

Has the PMO been established to serve a support and facilitation role? In this case, office staff may be composed of an experienced PMO director, a few professional associates with project management skills, and a clerical specialist to handle office functions. The chief function of this simple configuration may be to operate in ad hoc fashion to address project management issues within the organization as they arise.

If the project office acts simply as a repository of methodology and lessons learned, project mentors, methodology experts, or a project librarian may be sufficient as dedicated staff.

Some of the positions and titles we have seen within Strategic PMOs include: PMO director, project manager, project mentor, project controller, project planner, methodology expert, resource manager, metrics analyst, HR liaison, librarian/documentation specialist, relationship manager, administrative support coordinator, communications coordinator, issue resolution and change control coordinator, risk management coordinator, and software guru. For a summary of the responsibilities and skill requirements for several of these key roles, see Table 7.1. We have selected a few of the key roles for further discussion below.

Strategic PMO Director

If your organization is prepared to make the SPMO the central driving force behind the management of projects, you will want to consider establishing a director of project management who will sit at the director or vice president level with other senior executives in the organization. This position, which we will call the SPMO director, provides project oversight in virtually all areas of the organization, managing corporate level projects and overseeing corporatewide resource distribution and allocation on all projects. Any project that crosses divisional boundaries, as well as some large projects performed within a department, would be under the auspices of this SPMO director.

However, the SPMO director position is not merely a glorified project manager. He or she will have several critical roles to fill. The SPMO director must ensure that the project management process runs well while also seeking to continuously improve it. As the expert

Table 7.1 PMO Positions and Their Required Competencies

STRATEGIC PMO DIRECTOR	
RESPONSIBILITIES	SKILL REQUIREMENTS
Business interface	Leadership
Liaison to executive and functional management	Strategic planning
Development of standards, guidelines, policies, and procedures	Directing and managing programs
Project Management skills development	Building organizations
Resource prioritization	Identifying and developing new business
Project oversight	Selecting and developing key personnel
Project review and analysis	Multitiered management
Budget	
PROJECT MANAGERS	
RESPONSIBILITIES	SKILL REQUIREMENTS
Delivery of projects against schedule, cost, resource, scope, and quality baselines	Planning and allocation of resources
Resource management in schedule, staffing, budget, equipment, facilities, etc.	Leadership
Planning and control management to increase utilization and performance efficiency, reduce risks, identify changes, identify alternative solutions, etc.	Conflict resolution
Implementation of standard PM practices	Technical expertise
Interface management in product, project, client, and information flow, etc.	Team building
	Managing multidisciplinary tasks
	Entrepreneurial
	Administrative
	Focused vision
PROJECT MENTORS	
RESPONSIBILITIES	SKILL REQUIREMENTS
Senior advisor	PM competency
Counselor	Facilitation
Project visionary	Leadership
Facilitator	Team building
Planning direction	Conflict resolution
Support	Politics
Guidance to project teams	

Table 7.1 PMO Positions and Their Required Competencies (Continued)

PROJECT PLANNERS	
RESPONSIBILITIES	SKILL REQUIREMENTS
Schedule development	PM software
Planning support	Project software tools and methodologies
Resource forecasting	Cost estimating
Cost estimating	Office software use, such as word processing, spreadsheet, database, and presentation software
Critical path diagramming	Risk analysis
Software tools support	Change control
Project budgeting	
PROJECT CONTROLLERS	
RESPONSIBILITIES	SKILL REQUIREMENTS
Project integration	PM software
Resource forecasting and integration	Project software tools and methodologies
Budgeting	Cost estimating
Project and program reporting	Office software use, such as word processing, spreadsheet, database, and presentation software
Project prioritization	Risk analysis
Tools integration	Change control
Variance analysis	Database development
Reporting	Communication skills
METHODOLOGY EXPERTS	
RESPONSIBILITIES	SKILL REQUIREMENTS
Develop repository standards	Analytical
Develop training requirements on methods and processes	Logical
Develop performance guidelines	Methodical
Author, maintain, and update methods and processes	Organized
Evaluate, select, and maintain process management tools	Expert in methods and processes
LIBRARIAN/DOCUMENTATION SPECIALIST	
RESPONSIBILITIES	SKILL REQUIREMENTS
Manage and coordinate documentation revisions and releases	Administrative
Maintain repository standards	Organization

(*continued*)

Table 7.1 PMO Positions and Their Required Competencies (Continued)

LIBRARIAN/DOCUMENTATION SPECIALIST	
RESPONSIBILITIES	SKILL REQUIREMENTS
Maintain and perform periodic archiving of project records	Data entry, word processing, and spreadsheet tools, skilled with Web-based programs for archiving
Develop templates	Communications
Generate special reports	

ADMINISTRATIVE SUPPORT COORDINATOR	
RESPONSIBILITIES	SKILL REQUIREMENTS
Provide back-office support	Administrative
Perform scheduling and calendar functions	Organizational
Prepare travel itineraries	Data entry, word processing, and spreadsheet tools
Distribute project status reports and prepare project review presentations	
Maintain PMO personnel contact list	
Maintain office supplies	
Maintain and report action items lists and status	

COMMUNICATIONS COORDINATOR	
RESPONSIBILITIES	SKILL REQUIREMENTS
Help develop enterprisewide and project communications plan	Oral and written communications
Determine communication strategies and medium for information delivery	Interpersonal
Interface with internal and external organizations for information delivery	Organization
Ensure timely delivery of all project statuses	Administrative
Determine audiences requiring communications	Knowledge of communications delivery instruments

ISSUE RESOLUTION AND CHANGE CONTROL COORDINATOR	
RESPONSIBILITIES	SKILL REQUIREMENTS
Develop and maintain issue resolution and change control processes	Administrative
Establish standards	Organization
Create and distribute open issues reports for all projects	Analytical
Develop, maintain, and update issues and change control database/log	Communications

Table 7.1 PMO Positions and Their Required Competencies (Continued)

ISSUE RESOLUTION AND CHANGE CONTROL COORDINATOR	
RESPONSIBILITIES	SKILL REQUIREMENTS
Facilitate issues and change control meetings	Facilitation
Help project managers prioritize change requests	Knowledge of legal and regulatory requirements
RISK MANAGEMENT COORDINATOR	
RESPONSIBILITIES	SKILL REQUIREMENTS
Identify project risk during project definition	Risk assessment
Qualify risk	Analytical
Quantify risk	Alternative solutions/negotiations
Identify impacts	Communications
Respond through prevention, mitigation, and contingency planning	Facilitation
Monitor schedule and cost variance	Knowledge of legal and regulatory requirements

on project management, the SPMO director also serves as an ad hoc consultant and advisor to project leaders and teams. The existence of an SPMO director guarantees a focus on the consistent use of the project management process throughout the organization.

Michael Hammer wrote of the "two flavors" of manager: one is a process manager who oversees a process end to end, with skills of performance management and work redesign, and one is an employee coach who supports and nurtures employees.[8] A good SPMO director must be both: the overseer and "owner" of project management methodology and a leading mentor to up-and-coming project talent within the organization.

The SPMO director must possess enough stature and respect throughout the organization to champion projects from start to finish, and to recommend canceling projects whose objectives either can't be met or are no longer valid. He or she must have the demonstrable backing of senior management, especially critical early in the transition to the SPMO structure. However, instituting an SPMO director alone is not enough to bring the organization into a mode of "managing by projects." It is also necessary to alter the role of functional managers from resource owners to project resource suppliers, an equally significant change that organizations must make to fully realize the value of effective, cross-organizational project teams.

It's useful perhaps to think of the SPMO director in a similar light as a "program manager." Like a program manager, the SPMO director is responsible for the on-time delivery of projects that fall within the domain of the total program and is responsible for moving resources between and among projects as well as prioritizing projects within the program. While individually accountable for his/her own program (overall direction of the SPMO in our case), the director also may be measured and held accountable for the success of *all* projects within his or her domain. Thus, the structure supports the natural desire to work closely with other program managers (or, in the case of the SPMO director, with other division or departmental heads) on leveraging opportunities.

The following list includes just some of the many hats that may be worn by an SPMO director, or performed by personnel reporting to the director, depending on the size and scope of the SPMO:

- Relationship manager who works to smooth the interfaces with the business units, and develops project requirements through consensus with customers.[9]
- Communicating the mission, vision, scope, and benefits of the SPMO.
- Interfacing with all aspects of the business to increase a level of awareness of the services provided by the SPMO and the benefits of using those services.
- Serving as a liaison to executive and functional management.
- Developing the skills of the SPMO staff and of project managers throughout the organization.
- Prioritizing the application of PMO resources.
- Providing corporate project oversight, checkpoints, and controls.
- Reviewing and analyzing the process of project management throughout the organization.
- Managing the PMO budget.

As "owner" of the project office methodologies, the director also may be in charge of the following areas, either by taking personal responsibility for these items or by employing a methodology expert to fulfill these functions:

- Authoring, maintaining, and adapting the project management methods and processes
- Evaluating and selecting project management tools
- Contributing to definition of training requirements on corporate project management training, and project management methods and processes
- Developing knowledge management standards and processes for archiving and disseminating project documents, lessons learned, and other intellectual capital derived from project activities
- Developing tools for measuring the level of usage and effectiveness of project management methods used by the organization
- Soliciting and incorporating feedback from project managers for the continuous improvement of the methods and processes
- Defining and conducting project audits

The director also may act as a professional development coordinator—or oversee such a position—in order to ensure that:

- Job descriptions are created, maintained, and refined for the project management career path.
- Criteria are defined for interviewing, rating, and hiring for project management skills, as well as for identifying people in the organization who are currently acting as project managers and those who are interested in developing their skills in order to become project managers.
- Working with each individual to identify strengths and weaknesses in the project management discipline and assisting them in identifying opportunities for developing the appropriate skills and knowledge.
- Working with the internal training organization to identify project management training courses (or, increasingly, overseeing a training program managed from within the PMO).
- Working with management to identify individuals or assignment opportunities in order to develop skills and experience.[10]

In short, the SPMO director is an integrator of process, a manager of staff, a coordinator of project resources (including project managers and support staff), the coordinator of standards and methods as well as developer and maintainer of tools expertise, a mentor, training coordinator, and point of interface between projects, programs, and the executive staff.

A tall order, and one that must be filled with the same care that companies take in placing a CIO, a CFO, or, even, a CEO. In fact, over the past decade, we have seen the emergence of the "Chief Project Officer," on a peer level with other executives in the organization.[11] The alternative is to spread the project management-related responsibilities out among existing executives in the organization, which is too cumbersome to satisfy today's rigorous time-to-market needs.

Project and Program Managers

The goal of a project manager is to see to the successful completion of projects. This includes initiation, planning, execution, control, communication, and closure of the project. The project manager is also responsible for keeping the project sponsor and SPMO management apprised of progress and pertinent information. Coordination and communication with functional management and PMO team peers is essential as well. In addition, project managers today should be expected to feel a sense of partnership with the business, not just "do the job." A proactive project manager does more than merely report problems, he or she is empowered to take action, solve problems, and escalate issues to management as a last resort.

Normally, in a single-project environment, the project manager manages the day-to-day tasks necessary to move the project through all its phases, while keeping the project sponsor (for an individual project) apprised of progress and pertinent new information. As a member of a complex corporate SPMO, a project manager carries the same responsibilities with regard to individual project performance, but is somewhat relieved of the duties involved in keeping executive staff informed because the SPMO director serves as liaison between projects and the executive. The major exception to this general rule occurs when a project is in trouble. In a "troubled project," the PM is "invited" to discuss the project with the PMO steering committee, including what went wrong, the projected impact on the baseline, and corrective action planned. Naturally, these opportunities for exposure are not always welcome.

While communication and facilitation skills are always important for any level of successful project management, project managers, as part of the Strategic PMO and working within the formal PMO infrastructure and corporate project management methodology, are relieved of process and methodology design to focus on the technical and leadership tasks associated with their individual project success, such as project planning, time and cost estimating, leading the project team, controlling variance to the baseline, and successfully transitioning the project to operations and support.

All project managers are not created equal. Many have knowledge, skills, and personalities ideally suited to the project environment, while others do not. One of the stumbling blocks for project management growth in organizations in the past has been the failure to recognize that technical excellence does not translate to successful project management. Project management is a profession—a *discipline* that requires specific knowledge and skills in order to succeed at bringing projects in on time, on budget, and within specifications. Team management, negotiation, financial and business acumen, an understanding of organizational politics—all these areas and many more are required to demonstrate success in managing projects. The more competent the project manager is in these areas, the more capable he or she is to help the organization as a whole execute projects that are closely coupled to corporate and departmental strategic goals.

Project Support

Several roles complement the project manager to efficiently execute programs and projects:

- Project schedulers
- Project planners
- Project controllers

These roles, which represent the "science" side of project management, provide a career path for technically skilled project personnel, and by assuming responsibility for the tracking of schedules and cost, updating schedules and budgets, keeping project status reports current, and analyzing variances, they make sure the project managers and executives have accurate information upon which to base business decisions, and free up the project manager to concentrate on the facilitative and business aspects of projects.

Project Teams

Team members on projects make it all happen by executing the tasks necessary to move a project through all its phases to a successful closure and delivery. Typically, the number of team members increases as the project progresses and decreases as the project approaches delivery and closure. Full-time dedication to the team is also more prevalent during the execution stage.

Among the roles that will be filled by project team members in an SPMO include:

- Administrative support: back-office tasks, report generation, software support, calendars.
- Best practice or process experts: training, project oversight, quality assurance, methodology development, metric analysis.
- Librarian: project records, standards, methods, and lessons learned that must be stored in a project database. In a large organization, the maintenance of such a repository can develop to become a full-time job. Once envisioned as a clerical task, the SPMO librarian is now evolving into a sophisticated knowledge-management function and will become a fruitful source of benefits and value to the entire organization for historical data, successful practices, and effective templates, with knowledge that was previously lost with changes in and transitions of personnel.
- Resource Manager: Most organizations contemplating a Strategic PMO will want to maintain a resources database. The resources database is an inventory of all available resources throughout the organization. To complicate matters, the specific individuals available for assignment to projects constantly change as people join and leave the organization, technical resource skills are added and developed, people are

assigned to other projects, or individuals become otherwise unavailable. In organizations with significant project activity, the responsibility for resource management may become a full-time job. Individual project managers, rather than having to "beg, borrow, and steal" resources wherever they can find them, turn to the SPMO resource manager (RM) for assistance. The RM prioritizes resource requests and works with the PMO steering committee to manage the "fit" of resource skills to project requirements, manage and balance scarce technical resources, forecast and aid in planning for acquisition of resource shortfalls, and secure assignment of key resources to projects according to the project's relative rank on the organization's prioritized project list. Of course, all projects on the list are, by definition, linked to corporate strategy and each possess some degree of importance. If not of importance to the organization, projects should be canceled or rescheduled for execution in the future.

Other Team Members

The stakeholders in a project are those individuals and organizations who are involved in or may be affected by project activities. Key stakeholders, then, are the individuals (generally no more than eight to ten, and often only two to three) or organizations that are *most* affected by the project, and may include external customers, vendors, or regulatory personnel, depending on the project and industry.

Studies of organizational dynamics have revealed that a core team should number no more than eight people with five to seven being the ideal number. A study at 3M showed that team members should be located near each other; as little as 100 yards of separation severely hindered team interactions. Procter and Gamble, a company renowned for rapid product development practices, provides employees with desks on wheel to facilitate easy relocation of people when teams change.[12] That's because strong successful teams remain intact as much as possible. Keeping teams together is practical, especially given the time and effort needed to create a team from a new group of people; another argument for centralizing project personnel within an SPMO.

Most of these roles fall under the Project Support and/or Standards and Methodologies areas of PMO functions noted earlier. And, although we discuss these roles at the SPMO level, they are scalable to divisional PMOs as well. (Note: All these roles, and more, are fully described in our book, *Optimizing Human Capital with a Strategic PMO* (Auerbach, 2005).)

In addition, the new SPMO recognizes that people do projects and that most project problems are people problems. Having a "manager of project managers (MPM)," a role that combines project management experience with the skills to supervise, coach, mentor, develop, and recruit top project management talent, provides a long-needed interface between corporate HR and real life in the projects. One major insurance company includes a role called the "organizational development specialist" along with the MPM, whose role the IT PMO director calls "to float around and make sure everyone's human needs are being met," an addition to project life that has had tremendous positive effects on morale and productivity.

Then there's the "methodology guru (MG)." One of the complaints we have heard about PMOs is that they tend to become "methodology police." But, what we have in mind for the MG is more like a combination guru, cheerleader, and research whiz. Someone who is both passionate and meticulous and who can build, within the SPMO, a community of practice dedicated to changing work habits and project results through the application of a standard methodology. Imposed methodology is rarely sustainable, people being as contrary as they are. But, when introduced by someone fascinated with process and dedicated to ease-of-use and knowledge transfer, a methodology can transform a chaotic workplace into a rewarding one.

A Note on Mentoring

Providing expert advice and practical help in the form of mentoring to project managers is one of the core functions of a PMO. Mentors from the SPMO spend much of their time helping others. If the mentoring assignment is a large one, the SPMO will field a team of experts, organized under a team leader in a fashion similar to the organizational style of external consultants brought into the firm. The team leader is responsible for the scheduling and workload of the professional staff and the production of quality work.

Mentoring work is classified in six general areas:

1. Project mentoring
2. Project management
3. Project planning
4. Project administration
5. Professional development
6. Project audits

Project Mentoring

Project mentoring is the process of providing project management expertise to both internal and external project managers and project teams. Each individual is expected to maintain a high level of proficiency and knowledge in the areas assigned and to display that knowledge and proficiency during engagements and project reviews. Tasks will be defined for each project. Project mentors are not to be confused with project managers, who will have the full responsibility for project planning and execution. Project mentors may be assigned to individual projects or groups of projects, depending on project size and complexity.

Competency Identification

Research tells us that project goal achievement is influenced by four basic groups of factors: the superior project manager, the project office organization, the host organization, and the external environment. For purposes of generalized discussion, the project manager influences approximately 50 percent of project success, the project office organization about 20 percent, the host organization 20 percent, and the external environment 10 percent.[13]

As the component with the most influence on the probability of project success, a superior project manager has the ability to overcome nearly any *controllable* obstacle, and research indicates that the dominant events related to project success are generally in the controllable category. The project manager also is the key factor in recognizing and mitigating the impact of *uncontrollable* events. The project office organization might be nonexistent, the host organization could be weak, and adverse conditions might be encountered in the external environment. Nevertheless, the superior project manager will minimize these obstacles and work to achieve project goals.

This is most obvious when a project is in trouble. A responsive action taken by many organizations is to install a superior project manager. The broad competency groupings that compose the superior project manager are: (1) character, traits, and background; (2) professionalism consisting of leadership and management skills; and (3) project-specific skills comprised of the application of structured methodologies and procedures. According to the standards identified by the Top 500 Project Management Benchmarking Forum relative to project manager character, traits, and background, the best practice project manager:

- Is recognized by stakeholders as the single most important factor in project goal achievement
- Is truthful in all dealings and relationships
- Has a four-year college degree
- Is PMP® (Project Management Professional) certified
- Exhibits eagerness to organize and lead groups
- Exhibits evidence of a strong desire for goal achievement (degrees, certifications, ranks, and other goals achieved)
- Has above-average intelligence
- Is even-tempered
- Has faith that the future will have a positive outcome
- Has confidence his or her personal performance will result in a positive outcome[14]

Obviously, the SPMO has a stake in hiring, identifying, promoting, and rewarding the most competent project managers. In addition, as the center of project management within the organization, the SPMO can serve as an incubator of future project talent. The other areas of the organization will look to an SPMO to both provide excellent project management mentors and to develop their staff into superior project personnel.

In seeking to identify the competencies that should be required in a company's project managers and team members, the SPMO has a wide range of research on which to draw. Work by Lynn Crawford, which contributed to the Australian government's office competency requirements for project managers, is widely available in project management journals and on the Web. Since our first publication, the Project Management Institute (PMI) also has published its competency model. In addition, the PM College, working with the Caliper organization, has developed a competency model (described in detail in our 2005 book, *Optimizing Human Capital with a Strategic PMO*).

The establishment of a standard internal competency checklist for your organization is a critical first step for an SPMO. Without internal guidelines as to the skills and competencies required of project personnel, none of the most critical human resource issues, from hiring to career development to termination, can be equitably resolved. An example checklist for the position of project mentor is shown in Table 7.2 (Project Mentor's Competency Checklist).[15,16]

Training and Mentoring

Once you have identified the competencies desired for the various positions within the PMO, the next step is to ensure the availability of the necessary experiential and educational opportunities. To be effective, information collected from assessments of the knowledge, behavior, and aptitude of staff members is necessary to create a targeted training program. The educational program should be targeted to the requirements identified in the career path, and be designed in a progressive nature. In other words, the training requirements of team members are prerequisites for project managers and so on. Mentoring and coaching, in combination with other training and development, can help prepare the next generation of project leaders.

Our *Value of Project Management Training* study revealed that organizations overwhelmingly improved in a number of areas as a result of project management training. The relationship between classroom and workplace performance is highlighted by the finding that 91 percent of the organizations showed a moderate to extreme improvement in

Table 7.2 A Project Mentor's Competency Checklist

	NEEDS IMPROVEMENT				EXCELLENT
COMPETENCIES	**1**	**2**	**3**	**4**	**5**
Integrator					
Educator					
Expeditor					
Coach					
Problem Solver					
Quality Manager					
Risk Taker/Risk Manager					
Conflict Manager					
Partnering					
Visionary					
Information Powerful					
Flexible					

SKILLS REQUIRED	**1**	**2**	**3**	**4**	**5**
Facilitation					
Listening					
Team Building					
Negotiation					
Coaching					
Presentation					
Interpersonal					
Communication					
Conflict Management					

Note: The above is a list of competencies and skills required to deliver the highest quality project management mentoring expertise to our client. Take a minute and rate yourself on each competency and skill and then establish a personal development plan for improving those areas that need improvement.

the individual's on-the-job performance. They also show moderate to extreme improvement in a variety of business measures, including customer satisfaction, productivity, and cost schedule requirements performance.[17] Today, fully 70 percent of PMOs have training goals for their staff. Our *State of the PMO* study validated that training has an impact on organizational performance: PMOs in high-performing organizations are far more likely than those in low-performing organizations to have training goals (79.7 vs. 60.6 percent, respectively), and PMO staff in high-performing organizations are far more likely to be trained (70 vs. 50 percent).[18] And, in 2009, additional research showed that top performers spent three times as much on training as low-performing companies in the study.[19]

Who Should Be Trained? There are four audiences for training and each one needs a different depth and focus.

1. *Executives* need awareness training to give them an appreciation of what project management can do for them, and to teach them how to ask the right questions to get the right information so that they can make good decisions.
2. *Practitioners.* Project managers need training designed around their present level of capability. You don't design a uniform training program, you design a program that meets the needs of the population. Train all levels, from novices to experts who need refreshers. Assess your practitioner audience for areas of strength and weakness and design around that.
3. *Matrix organization functional managers* need training in three areas: their role in a project environment (as resource owners), the role of project management in the organization, and, finally, an overview of what project management is all about. A related audience is ancillary managers, such as HR and finance, who support projects. The need for functional managers to receive project management training was highlighted in a recent field study published in the project management journal. One of the findings was that, even when project managers are knowledgeable about risk management, poor risk management results when functional managers do not have any formal risk management training.
4. *Team members* on the project need to understand both the big picture and their supporting roles. Encourage "would be" project managers to participate in the training as well. Use training to grow the skills you will need to address the staff transition that will occur, to ensure you have a pool of trained staff to draw upon.

Executives and project managers should be trained simultaneously. The executives are change agents and project champions. If project managers know that executives are going to be asking them questions that they might not be able to answer, they pay more attention. It improves performance at the same time it raises expectations on the part of executives, so the two trainings reinforce each other.

Most individual workers believe they need training. Most corporate executives also believe training is a high priority. But often, middle managers need to be educated on its importance. Managers send a strong message about the value of employees via policies on who pays and when the training is delivered. When management wants employees to pay for the class and use vacation time to attend, the messages that employee receives about trust, power, and career are not positive.

Reinforce the training with follow-on activities to maximize the effectiveness of the training. Implement a mentoring program that supports project managers benefiting from the experiences of other who have gone through the training. Ask your instructors to conduct follow-up sessions; bring the entire class back together to talk about their experiences applying the classroom instruction in the real world.

Companies can support individuals—and improve training results—when they are mindful that individuals are more likely to take ownership for their learning when:

1. They are given a compelling, meaningful reason to do the task (i.e., a link to strategic goals).
2. They have options to make the task more interesting.
3. Social networks exist to support the learning (so that individuals fulfill social needs as they connect with one another through a topic or project of interest.) The project management community of practice is an example of such a social network, as are online threaded discussions with other practitioners, professional associations, and the like.

Mentoring and Coaching Mentoring is the best source to better understand work standards and behaviors that fit the company culture. Effective mentoring involves a relationship with a respected executive-level person to gain their evaluation of your professional and personal goals, their insight into the development of action plans, and their experience on how to develop relationships or handle specific situations. Mentoring is the most effective way to bring new project managers up to speed quickly. Why? Mostly because practical experience is the best teacher. Once the basics are learned through training, project management expertise is gained from on-the-job training. Experience, under the tutelage of a mentor, is the best teacher.

Performance Measurement and Rewards

Individual performance objectives are generally arbitrary because they cannot account for the interdependence of the team's tasks. As we generally use them, individual performance goals are simply subgoals based on the functional process measures of more senior managers. What is surprising is that many companies are just beginning to realize that the ways performance is measured and compensated determines whether or not teams reach their full potential.

The most radical, yet most obvious, change is to base all performance appraisals and review systems on the team, and make the team accountable for team results. For true teamwork to occur, people need common purposes, measurable goals, and a common fate. Thus, moving toward a project-oriented organization means creating a team-oriented appraisal and reward system. Because the team is in the best position to control the task, the team should be the primary focus of any performance measurement. Functional expertise is very much a prerequisite to team participation, but it is appraising performance based on the team's results that encourages people to wear two hats.

By measuring performance in the context of the whole process, we can begin to overcome the functional silo mindset, which encourages people to focus on their function to the exclusion of the project customer. In the project-oriented company, all employees take responsibility for interpreting the voice of the customer and acting on that to feed new ideas back into the system.

Individual contributors to projects will ultimately benefit from the institution of an SPMO, but there also will be changes and dislocations for them to adjust to as they give up certain tasks and take on responsibility and authority to meet customer needs, while gaining the required expertise in their function.

Glenn M. Parker of human resource consulting and research giant Watson Wyatt recommends reward structures that foster collaboration, in which individuals are acknowledged, but primarily for being strong team players; those who "help the crowd stand out, rather than standing out from the crowd."[20] An example of a simple assessment of project management service delivery quality, in this case for a Project Mentor, is shown in Table 7.3.

Table 7.3 Mentor Quality of Service Delivery

	NEEDS IMPROVEMENT				EXCELLENT
DOES YOUR PROJECT MENTOR:	**1**	**2**	**3**	**4**	**5**
Seek out your requirements, priorities, and expectations?					
Effectively support your project and/or program?					
Treat you with respect and understand your situation?					
Solicit, listen to, and resolve your concerns?					
Provide timely advice?					
Deliver quality information and guidance?					
Display flexibility in responding to your needs?					
Keep you informed?					
Would you select this mentor for future projects?					

Note: In performing as a mentor, ask yourself the following: "What would be my client's response if asked these questions in regard to me?" Then rate yourself again to determine where you need to improve.

RATE YOUR MENTOR ON THE FOLLOWING:	**1**	**2**	**3**	**4**	**5**
Project Management					
Project Planning					
Project Scheduling					
Communication Skills					
Interpersonal Skills					
Judgment					
Team Participation					

Note: If your client is asked about your knowledge, skills, and abilities in the following areas, what would be the response?

Career Paths and Leadership Development

Once the competency checklist is established, the SPMO works closely with an organization's HR department to develop position descriptions and grades and to establish a career path and promotion path combining project management assignments, years in the profession, training, and certification.

Recent job satisfaction research has revealed that one of the primary reasons project managers leave a company is the lack of a clear career path. Formerly, the only way project managers, who were primarily technical staff, could rise in the organization was by leaving their technical specialties and project management skills behind to climb the supervisory and administrative ladder, positions which were often a poor fit with the project manager's temperament, abilities, and talents. With the evolution of the PMO, career project managers can now look forward to managing projects, managing other project managers, directing a PMO, participating on the steering committee, and even, in some cases, rising to corporate executive status based on the performance of the projects under their care. However, this path must be incorporated into the company HR documentation and made known to project managers and others in the organization or it remains only a possibility, not a policy. See Figure 7.1 for an example of the professional development path for project managers, based on a program developed by the PM College.

A broad range of experiences are required for future project managers. It is not possible to develop them by restricting their experiences to one function. Thus, rather than climbing the ladder up the functional silo, project managers benefit from being exposed to a number of functions, perhaps moving back to functions they have fulfilled before, but in a more senior role. It is far better, however, if alternative upward paths exist: one through technical managership and one through project managership. With such dual promotional ladders, technical managers can stay in their departments and become core team members responsible for the technical portions of projects. Dual ladders also allow progression through project management, but project managers must be able to manage technical specialists while handling the behavioral and administrative tasks that motivate the specialists to do their best work.

Position / Organizational Level
Project Management Business-side Career

Chief Project Officer / Enterprise
Strategic Project Office Director / Enterprise
Project Office Director/ Divisional / Departmental
Portfolio Manager / Enterprise
Manager of Enterprise Project Managers / Enterprise
Manager, Project Managers / Divisional/Departmental
Global Program Manager / Enterprise
Enterprise Program Manager II / Enterprise
Enterprise Program Manager I / Enterprise
Enterprise Project Manager II / Enterprise
Enterprise Project Manger I / Enterprise
Program/Project Mentor I, II / Enterprise or Divisional
Program Manager II / Divisional
Program Manager I / Divisional
Project Manager III / Divisional/Departmental
Project Manager II / Divisional/Departmental
Project Manager I / Divisional/Departmental
Business Analyst
Risk Management Coordinator

Project Planning and Controls Focused Career

Chief Technology Officer or CIO / Enterprise
Strategic Project Office Director / Enterprise
Project Office Director / Divisional/Departmental
Manager of Enterprise Project Support / Enterprise
Manager, Project Support / Divisional / Departmental
Enterprise Project Controller / Enterprise
Project Controller II / Divisional / Departmental
Project Controller I / Divisional/Departmental
Project Planner II / Divisional / Departmental
Project Planner I / Divisional / Departmental
Project Estimator II / Divisional / Departmental
Project Estimator I / Divisional / Departmental
Project Scheduler II / Divisional / Departmental
AProject Scheduler I / Divisional / Departmental
Business Analyst
Issues Management Coordinator
Change Control Coordinator
Risk Management Coordinator

Specialty Positions / Careers

Systems Analyst
Knowledge Management Coordinator
Methodologist
Technical Advisor
Budget Analyst
Project HR Specialist

Figure 7.1 An example of the professional development path for project managers, based on a program developed by the PM College.

A project manager multilevel career path can be developed around levels of competence in the following categories: education/experience, interpersonal skills, technical coaching ability, business acumen, customer focus, project complexity, role in managing projects, and the industry/work environment.

According to Christopher Sauer of Oxford University, developing a career structure is essential to the development of an organization's project management capability. The career path structure serves three purposes:

1. It allows the organization to match a project manager's level of competence/experience to the difficulty and importance of a project.
2. It assures project managers that the investments they make in developing their professional skills will be rewarded.
3. It provides an incentive for people to stay with the company, because they can see a clear promotion path.

To survive in the competitive business environment, an organization must develop a career program that will respond to future strategic requirements for certain skills. By doing so, the organization provides opportunities for and encourages multifunctional experiences and allows their personnel choices and control over their lives. To the extent that the organization provides opportunities for the individual to use and develop his or her personal competence while moving through various jobs, functions, and levels, the individual will grow and experience career satisfaction, and the organization will reap the benefits.

The objectives of any career development program should be to:

- Improve skills
- Assess an employee's readiness for advancement
- Define professional skill areas
- Create an equitable salary structure
- Create a positive and open environment for career discussion
- Ensure frequent feedback
- Provide opportunity for advancement through a career "path"
- Encourage a "change to grow" environment
- Assure future strategic leaders for the organization

The job descriptions, competency analyses, and career paths recommended for the SPMO are all geared toward creating a professional development environment focused on maximizing the talents of all project management personnel. Supported by these processes and artifacts, managers can focus on opportunities that support the individuals' development of their talents while, at the same time, adding to the skills and knowledge of that individual. When professional development is directed at getting better at what we do best, company performance can only improve.

Best Practices for People Management in the Strategic PMO

Having a Strategic PMO involved in training, managing, measuring, rewarding, and promoting project people is a key factor in effective strategy execution. Best practices identified for "people management" in our *Strategy & Projects* research included:

- Project stakeholders understand how they can influence the successful execution of strategy and how their work is important to execution outcomes.
- Project stakeholders have clearly defined individual and team performance targets that are aligned with strategic objectives.
- Performance management reviews are structured to reward or correct individual performance based on the employee's contribution to strategic objectives.
- Project stakeholders clearly understand and buy in to the organization's strategies.
- The project management staff is capable of creating, deploying, and maintaining enterprise, portfolio, program, and project strategies.[21]

The most often used practice by high-performing organizations is having project stakeholders buy in to the organization's strategies. High-performing organizations are significantly better than average at having performance management reviews structured to reward or correct individual performance based on the employee's contribution to strategic objectives.

Rolling Out the Strategic PMO: All Aboard

Projects cannot be accomplished by just a project manager and team. The individuals "at the coal face" of a project cannot be held responsible for many of the potential obstacles to success, such as poor definition of the business case for a project, lack of alignment with strategic corporate objectives, inadequate funding, or refusal to cooperate cross-functionally among the departments. These types of obstacles must be cleared away at the executive level of the organization; that is why the project sponsor plays such a critical role in the successful management of an individual project.

When initiating an organizational change project on the scope of a PMO, executive sponsorship is even more necessary. The level of authority required to drive this kind of change does not exist on the project team or even the departmental level. Only the company's executive staff has that authority and responsibility to design and oversee the implementation of this magnitude of change. The executive leadership of the enterprise must commit the organization to this new direction and exhibit the resolve necessary to see these changes through to completion. Many sources in the literature on successful reengineering and implementation of new processes agree that executive commitment is the first and most crucial piece in any drive to improve or change organizations. Thus, in discussing the staffing of the Strategic PMO, we will "take it from the top."

The Executive Role

The primary role of the executive staff is to provide the strong leadership, strategic vision, and program definitions necessary to implement the PMO. A company can have a best-in-class project management process defined, but if the strategic vision that underpins that process is missing or ill-conceived, the process simply cannot make that company successful over the long term.

The executive staff must establish vision and direction for the project management initiative and allocate funding and resources to it. Such sponsorship from the members of the executive committee ensures a voice for programs and projects. Many organizations have strong support for project management at lower levels, but very little acceptance or interest at the top. Such an organization is not managing effectively by projects and is unlikely to derive the benefits that enterprise project management has to offer, unless a cultural evolution takes place. It isn't necessary for executives to become project managers; however, it is necessary that they enthusiastically support, with words, actions, funding, and support the aspirations of the project management community within their organizations.

Once the initiative is in place and the projects that fall under it are gearing up, the executive staff has minimal involvement in day-to-day project activities. However, even though the executive team is not performing the day-to-day detailed work, they must be involved as an executive oversight team. Ideally, management understands the strategic implications of the PMO initiative and its impact on the company's bottom line in terms of more rapid new product development and the resulting increased return to shareholders.

Where Executives Fail The dilemma that arises in many organizations is that the impetus toward improved project management begins at middle management level. And, because middle management tries to implement changes that are really beyond their scope of influence, the project management initiative usually receives inadequate funding. Middle management usually only receives enough support to cobble something together—*if* they didn't get tied up with other things. Without executive management sponsoring/driving/overseeing PMO development, project management initiatives do not deliver the promised value, not because the initiative itself was not a sound idea, but because the implementation is half-hearted. Without the backing of executive management, project office resources typically find themselves implementing the project office initiative as a part-time role, are periodically pulled from deployment efforts to manage current "hot" projects, and find themselves juggling multiple priorities, many times defocused to the point the project office initiative loses is direction and momentum.

Make no mistake, deploying a corporate-level PMO is a strategic program. Thus, management has a critical role to play.

How to Succeed What are the critical elements of successful management participation in a Strategic PMO deployment? There are many, including the strategic decision making that supports the project, protecting the resources in the SMPO, so that they can focus on what they need to do: supporting the budget, supporting the plan, supporting the schedule, and providing conflict resolution when resistance to change arises within the departments most affected by the deployment. However, the keystone in the deployment strategy is the executive sponsor. The executive sponsor paves the way for the deployment by dealing with other executives as a peer when conflicts over resources arise.

Without an executive sponsor, the chances of successfully deploying a PMO are very slim. The more influential the sponsor, the greater your likelihood of success.

Identifying the Executive Sponsor

In identifying an executive sponsor for your SPMO deployment project, it might be helpful to think about the usual role of the sponsor on the kind of individual projects with which you may be more familiar. The project sponsor is the executive in charge of the area in which most of the business functions connected with the project reside. He or she initiates the project and is a member of the oversight committee. The sponsor makes business decisions at the various project phases, communicates the larger vision of the project throughout the organization, and, from the customer's (the executive leadership team, in this case) point of view, is ultimately responsible for the project's completion. Project sponsorship is most effective when accountability resides with one person, a person high enough in the organization that he or she has enterprisewide influence. There will of course be a project sponsor for every project undertaken by the company, but the sponsor chosen to spearhead the PMO initiative will have a particularly important role in leading organizational change; thus, the sponsor of the PMO initiative must be highly placed.

For an SPMO deployment supporting the technology division, a chief information officer (CIO) could also serve as an executive sponsor, depending on the level of influence he or she wields within the organization. Some organizations have realized that IT is integral to overall corporate success and their CIOs have won a seat at the strategy table. A CIO with this kind of organizational clout can be an appropriate executive sponsor; however, in most companies a higher-level executive will be more effective; in some cases, this may even be the chief operations officer (COO) or even the chief executive officer (CEO). However, in our experience, in a midsized organization or within a business unit of a larger organization, the executive sponsor would typically be at the vice president level. As you move up into the Fortune 100, a junior vice president or director-level executive may serve as the executive sponsor. A simple rule of thumb for choosing a person with sufficient authority is simply this: Does he or she have the authority to cancel the project?

Lack of an executive sponsor with sufficient authority is a major risk to the success of your initiative, and we recommend that work not proceed until you engage an effective project office sponsor. This step should take place early in the initiative to ensure the project will move forward. Securing buy-in across the executive positions in your organization significantly improves the probability of project office success, giving the project team the ability to resolve the kinds of issues, conflicts, and challenges that occur whenever you try to deploy an organization-wide system of this magnitude.

Coordination of the project/program/organizational budgeting processes, procurement, inventory control, capital equipment funding and allocation, and suppliers—there is a tremendous amount of coordination and systems integration required for a fully functioning Strategic PMO. In order to expedite these integration issues, the executive sponsor must be a champion for the SPMO, while serving as an effective organizational facilitator. Being seen as a proponent of this process change for the good of the organization as a whole allows the sponsor to pave the way to work through some of the sticky issues of turf, information-sharing, and power.

The Bottom Line Choose a sponsor for the SPMO who can communicate the plan, and keep the organization's priorities straight. He or she must be a strong advocate for the changes involved, extremely knowledgeable about the benefits of project management, and have the ear and confidence of the powers-that-be. The old saying goes, "You're never a prophet in your own land." If senior management doesn't fully understand and support the project management approach, it may be time to bring in an external consultant who has dealt with a number of companies in your market segment to explain and execute the advantages of project management and the results achieved by others who have successfully implemented a PMO.

Management Participation: The PMO Steering Committee

As the liaison between senior corporate management and the SPMO project team, the executive sponsor should be the chair of the PMO steering committee. This committee is normally made up of the director of the SPMO, the project sponsor, the heads of key functional organizations (members of business units affected by the project or projects being dealt with at any one time), and a senior corporate official, such as the CEO or COO. I recommend the PMO steering committee be comprised of three to seven individuals total. This committee is formed to change the corporate project culture and is active on a continuing basis to select, prioritize, and evaluate the entire corporate portfolio of projects. In addition, it acts specifically on very large projects having overall corporate impact, such as the SPMO initiative. When major issues or problems with the project must be escalated, the PMO steering committee provides a forum for issue/problem resolution. This committee initiates the project in a management oversight role, and also continues to hold end-of-phase reviews throughout the duration of the deployment project, monitoring progress against the objectives to determine whether or not the SPMO is meeting the objectives that were established at initiation. The PMO Steering Committee also may discover the need to include technical and internal client representatives—senior staff from other business units that may be affected by the deployment. If there are external customers who are critically affected, you may want to include them on this committee as well. This group is, in effect, the board of directors for the SPMO and other megaprojects.

As a "board of directors," the PMO steering committee has input into the strategic direction and will play a part in the review of the SPMO charter. In some cases, members of the committee will need to sign off on key elements of the deployment plan (such as the project charter) because the charter defines the scope of the proposed SPMO and its specific roles and responsibilities with respect to functional departments and business units. While the SPMO director will write the initial draft of the charter for the PMO steering committee meeting, he or she will ask committee members to sign the charter to verify that its provisions have been agreed to. If a conflict arises in the future, the members of the committee revisit originally agreed-upon terms of scope, priorities, and strategy prior to initiating change. The PMO steering committee also will continue to revisit the goals and objectives of the SPMO, as well as the critical deliverables, and continue to work within the organization to achieve executive buy-in to all those areas.

The PMO steering committee is involved in the commitment of all the various resources that the SPMO deployment will require, from budget and personnel to space, equipment, and time.

In the early stages of an SPMO deployment, the PMO steering committee will be required to meet more frequently, perhaps as often as monthly. As the project begins to deploy, the committee will meet less often. As part of project planning, the PMO steering committee may wish to identify key end-of-phase points when they will come together to review progress to date, determine whether the objectives of that phase have been achieved, whether the schedule has been maintained, whether proper cost controls have been put into place, and so forth. Another way executive management can make sure that PMO steering committee representatives fully appreciate the importance of the project is to ensure that committee members devote sufficient time to committee proceedings; that they, in fact, attend meetings and provide meaningful input, and provide feedback to senior executives on progress and problems. It may be necessary to conduct a session of executive-level training in project management before the SPMO deployment project is launched to ensure the PMO steering committee fully understands its role, responsibilities, commitments, and value.

Functional managers seldom have a sufficient grasp of the enterprise advantages of project management to fully appreciate "what all the fuss is about." The members of the PMO steering committee must understand enough about the project management process and its value to be strong advocates; and they also must have a high-level understanding of the phases and processes of the discipline in order to provide leadership and guidance during project reviews. Periodic progress reviews are a normal part of the "controlling" processes of project management. On the SPMO deployment project, as on any other organizational project, executive participation will be necessary in these project reviews. These reviews will be scheduled into the implementation plan. One reason executive involvement is important in these reviews is that the PMO steering committee must have the authority to both launch and cancel the project if necessary. Otherwise, the committee will not have sufficient authority to make other critical decisions necessary for the successful outcome of the project. Project control is all about identifying problems, risks, or issues early in the deployment initiative, and addressing them as a project moves through it's life cycle of planning, deployment, and transition to ongoing operations.

Keep in mind, however, that the PMO steering committee must be aware of the highlights of the program only: a very high-level roll-up of all the project activity. The SPMO project manager should provide the committee with an agenda and a menu of decisions that must be made during the PMO steering committee meetings. Some of the typical issues the committee will be asked to address include major changes to the direction of the Strategic PMO deployment project or other significant change control items, budgetary impacts, resource conflicts, need for involvement from other organizations, or lack of support from a critical "power center" in the organization. A simple word from the CEO or COO to a recalcitrant player is sometimes all that is needed to get the PMO deployment back on track. All the issues the project team itself are not able to resolve should be elevated to the PMO steering committee so that the committee can use its influence or decision-making ability to redirect, correct, provide funding/resources, reprioritize, or take other action. And, as always, it is necessary to document committee decisions and incorporate them into an updated plan or issues log.

Lastly, one of the most important areas in which the PMO steering committee plays a role is in the realm of culture change. As we discussed in Chapter 1, managing by projects is an entirely new way of doing business in many organizations and anyone attempting to align projects and strategy will impact not only those individuals doing project management, but functional teams and managers, and systems from HR to payroll to facilities to procurement to finance. Changes of this magnitude cannot take place without management support and advocacy, and this will be a primary role for the PMO steering committee. It has been said that much of implementing project management is "missionary work" and the executives involved have to be the primary "missionaries" of this new business doctrine.

How to Select a PMO Steering Committee

The PMO steering committee should be made up of three to five executives in the organization. The chair of the steering committee must be the executive who stands to gain (or lose) the most from a successful implementation of a PMO. The remainder of the steering committee is represented by corporate leaders whose organizations maintain a vested interest in effective deployment of a project management culture. It is critical the PMO steering committee possess the authority to (1) approve the plan and budget for deploying a PMO, (2) authorize changes to the PMO deployment plan and budgets, (3) provide resources for the initiative, and (4) assist the PMO deployment manager in building support for the PMO, resolving conflicts, and successfully deploying a project management culture. The capstone consideration for qualifications to participate as a member of the PMO steering committee is that the steering committee must have the authority to cancel the PMO Deployment Project should conditions warrant.

Notes

1. J. Kent Crawford and Jeannette Cabanis-Brewin, *Optimizing Human Capital with a Strategic Project Office* (Boca Raton, FL: Auerbach Books, 2005).
2. James S. Pennypacker, *Resource Management Challenges: A Benchmark of Current Business Practices* (Glen Mills, PA: PM Solutions' Center for Business Practices, 2008).
3. Ibid, p. 10.
4. James S. Pennypacker, *The State of the PMO 2007–2008: A Benchmark of Current Business Practices* (Glen Mills, PA: PM Solutions' Center for Business Practices, 2008).
5. Ibid.
6. James S. Pennypacker, *Justifying the Value of Project Management* (Glen Mills, PA: PM Solutions' Center for Business Practices, 2001).
7. Christopher Sauer, "Where Project Managers Are Kings," *Project Management Journal* (December 2001).

8. Michael Hammer and James Champy, *Reengineering the Corporation* (New York: HarperCollins, 1993).
9. M. Light and T. Berg, "The Project Office: Teams, Processes and Tools," *Gartner Strategic Analysis Report* (August 1, 2000).
10. Carolyn M. Hennings, "Proposing a program Office for a Service Organization," paper presented at the *Proceedings of the 30*th *Annual Project Management Institute Seminars & Symposium* (Newtown Square, PA: PMI, 1999).
11. Paul Dinsmore, *Winning in Business with Enterprise Project Management* (Chanute, KS: AMACOM, 1998).
12. Brian Dumaine, "How Managers Can Succeed through Speed," *Fortune* (February 13, 1989).
13. Frank Toney, *The Superior Project Manager* (Glen Mills, PA: PM Solutions' Center for Business Practices, 2001).
14. Ibid.
15. Paul O. Gaddis, "The Project Manager," in *The Harvard Business Review* collection of articles *Project Management*, ed. President and Fellows of Harvard College (Cambridge, MA: Harvard Business Review Press, 1991).
16. Lynn Crawford and Fran Gaynor, "Assessing and Developing Project Management Competence," paper presented at the *Proceedings of the Annual Project Management Institute Seminars and Symposium* (Newtown Square, PA: PMI, 1999).
17. James S. Pennypacker, *The Value of Project Management Training* (Glen Mills, PA: PM Solutions' Center for Business Practices, 2006).
18. Ibid.
19. James S. Pennypacker, *The Value of Project Management* (Glen Mills, PA: PM Solutions' Center for Business Practices, 2009).
20. Glenn M. Parker, *Cross-Functional Teams* (San Francisco, CA: Jossey-Bass, 1994).
21. James S. Pennypacker, "Strategy & Projects: A Benchmark of Current Business Practices" (Glen Mills, PA: PM Solutions' Center for Business Practices, 2005).

8

The Technical Infrastructure

Using IT to Facilitate Project Collaboration and Performance Measurement

This is a chapter about tools that doesn't really talk about the tools themselves, but about the human and organizational structure in which they are embedded. In today's volatile software marketplace, there would be little point in the author of a book attempting to discuss the attributes of specific software products when these products may be obsolete by the time the book hits the shelves. That's one reason you won't find any software recommendations in this book. But, the other reason is that what software you use is far less critical to your success than how you use it, who uses it, and how you selected it.

A PMO needs three key elements to operate smoothly: people, process, and tools. Without the proper emphasis on each element, an organization will have difficulty making changes and improving the way they operate. You may have the best people, but without the right tools on hand, productivity will suffer. You may have the best tools, but without the right process all you have done is automate an otherwise bad situation and your people will become even more frustrated. You may have the most comprehensive process, but without the buy-in of the people in the organization, that process isn't going to change the work culture. You need all three elements to make changes and, in particular, changes and improvements in project management.

That said, let's focus briefly on software tools. We'll explore some key questions. What kinds of functionalities are important for the various levels of PMOs? What are the best practices for selection? After you've selected your software tool, how should you roll out the software within the organization? The objective of this chapter is not to tell you what kind of software to use; that is a complex question; that can only be determined through a rigorous selection process.

Instead, we will raise issues and questions that are important for any company implementing a PMO to consider.

Why Do You Need PMO Software?

This question may seem obvious. It's just common sense to buy a packaged tool to jumpstart the PMO, isn't it? Maybe. However, many companies jump into the tool-buying stage without proper consideration of their specific needs and requirements. We recommend you think through your requirements before talking with software vendors. To help you in this process, here are some thoughts on why you need software. Select the ones that fit your situation and put them in order of priority. This will help you to ultimately select the right package.

Project management software for the PMO must:

1. **Support integration for a total enterprise perspective**. In many cases, organizations are looking for software that will depict the big picture. You need the ability to summarize and capture the most pertinent information for an executive perspective. As things change in the supporting details, you want the top-level information to reflect the impacts and the new status. Not only do you want to see this for the enterprise schedule, but you want to see it for all your resources and project funding, too. Having the ability to efficiently integrate the details and provide a total enterprise perspective for schedule, resources, and cost is a capability of most sophisticated tools, but not of all tools. If this is a priority of yours, make it one in your selection.
2. **Provide different capabilities and levels of information for diverse stakeholders who work at different levels**. Executives are important, but we can't forget about the project managers and their teams. The software tool needs to be sophisticated enough to provide the enterprise integration capabilities, but it should be simple enough to use on an individual project. Don't buy a software package that is overly complex and not intuitive. Make sure it is user-friendly and easy to migrate through (get your experts to test it and tell you). Some software vendors have an integrated set of software tools targeted

at the different levels and needs of the software users. As an organization, you would get the simplicity for the single focus of a project team and the complexity for the enterprise focus of the organization—all easily integrated within one software suite. Make sure you think through the needs of all users.

3. **Support dynamic, changing environments and priorities**. Doesn't this describe almost every organization? Most entities are dependent upon external influencing factors. Priorities shift. Perhaps the R&D department made a major breakthrough and project funding is realigned to pay for the startup project. Or, perhaps another project is in dire straits and resources are pulled from three other projects to help. That's why you need a software package that is flexible and can easily adapt to changes across the board. You don't want to manually reconcile different databases every time there is a fluctuation. Get a tool that makes it easy to make changes, not hard. (See comments on friction later in this chapter.)
4. **Make a good process more efficient with software**. We have seen good status reporting processes that took forever to complete each month because most of the consolidating was done manually. Assuming your project teams plan and track progress within the software tool, it is much more efficient to have the tool summarize and provide total information for you to analyze. In this case, a good process definitely becomes more efficient with software automation. Select a tool that supports or enhances your processes. Identify the ones that are important and understand how the software can support them.
5. **Support geographically separated, virtual teams**. Nowadays, virtual project teams and PMOs are accepted as the norm. Today's technology makes it feasible to build project teams and PMOs with experts from across the company. This virtuality in an organization adds another layer of complexity to communications management. E-mail, the external Web, corporate intranets, and the good, old-fashioned telephone are communication instruments that support virtual teams. If your structure is virtual (or your technology architecture is heading that way), you need to select a software package that will operate in such an environment and be accessible by

all parties regardless of their physical location. Additionally, social media tools now play an increasingly important role in project teambuilding and collaboration, especially for distributed teams.

6. **Provide a historical database to gather company-specific information for future planning and estimating.** This is definitely a planner's dream. Can you imagine actually having historical information to estimate how long a task will take and how many resources you need to accomplish it? Imagine having access to how much it cost on similar projects to get the same effort done. A software tool in the PMO can be a central repository of information and collect actuals on projects as they move through the life cycle. If this is important to you, understand how this is accomplished and maintained within the software you are selecting.
7. **Reduce the administrative burdens.** Let project managers spend more time managing and analyzing rather than consolidating and summarizing those infamous status reports. Take a hard look at the software's capability to produce reports, and, in particular, the ease to tailor and develop your own reports. You want to minimize the time spent by project managers in consolidating and summarizing information. You want to maximize the time they spend on managing the project and interpreting the resulting information. Make sure the software maintains and provides the information you need to see.
8. **Support the establishment of standards and consistency across projects.** In large part, software tools will support standardization and consistency. By their very nature, they are oriented in this direction with common database fields and nomenclature. In many instances, the software is simply a shell for you to populate as it best fits your organization. If you are big into templates and tutorials, you should ask about the software's capability to support such things.
9. **Provide a central place for written communications and project exchanges.** Most folks like to keep things simple and are creatures of habit. It is convenient to have one place, a central place, to find stuff. It is a good idea to centrally locate project management guidance and new "things happening in

the field." Individuals then will know where information is located and can access elements that will help their current activities. Ask the software vendor if their package provides a central location (or repository) to organize your information.

10. **Manage the project portfolio and provide high-level executive dashboards of pertinent strategic project metrics**. The role of PPM software in optimizing existing resources cannot be stressed too firmly. In our 2008 resource management research, it was the organizations without mature PPM processes that struggled most to identify, allocate, and manage their human resources.[1]

Software Functionality and PMO Complexity

We can think about what type of software products the PMO needs in two ways: by project management approach and by the level of the PMO in the organization.

Project Management Approach

There are two basic approaches to project management that are reflected in software products:

1. **Task management**. The project is seen largely as a collection of tasks to be accomplished. Organizations using this approach assume that the resources will be available when needed to accomplish the specified tasks. This assumption may be valid when resources are plentiful and the number of projects is low, lessening the chance for interproject conflicts. Low-end desktop tools (even spreadsheet software) can be used to manage projects under this approach. Lower-end tools suitable for this type of project management are easy to implement and relatively inexpensive. Unfortunately, few organizations today have a light workload of this type. Tools based on this paradigm generally constrict a PMO's ability to plan, organize, and control multiple, complex projects in an enterprise portfolio. Scheduling tools are "necessary but not sufficient" for managing a portfolio of projects. A multiproject view of work

is required, with facilities to ease collaboration, automate skill and resource management, facilitate issue management, and assist with metrics collection and organizational learning

2. **Resource management**. A primary constraint on successful completion of projects is availability of key resources. This approach requires more sophisticated methods and tools to gather information about project status and to reconcile resource requirements across projects; without sufficient investment in these methods and tools, little improvement in capability will be possible. Higher-end tools support multiproject management via resource sharing and leveling functionalities. While most software packages include these functions, they are underutilized.

Since most organizations are complex enough to have both simple and complex project activities in the works, Gartner, Inc. studies show that most enterprises require various classes of project management tools to support business requirements, along with the training to use them appropriately. [2]

PMO Level

We've discussed the three levels of complexity in a PMO, ranging from single-project endeavors to the Strategic PMO. In considering what kind of software is needed, the complexity of the PMO is a starting point. Figure 8.1 shows how the need for software functionality co-evolves with the project management practice in an organization.

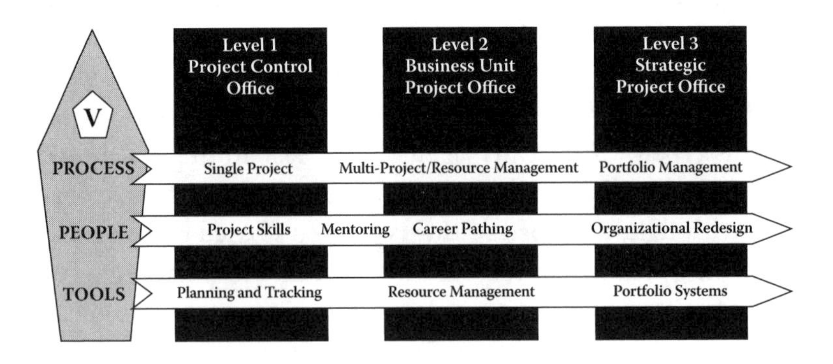

Figure 8.1 Tools in the evolving organization.

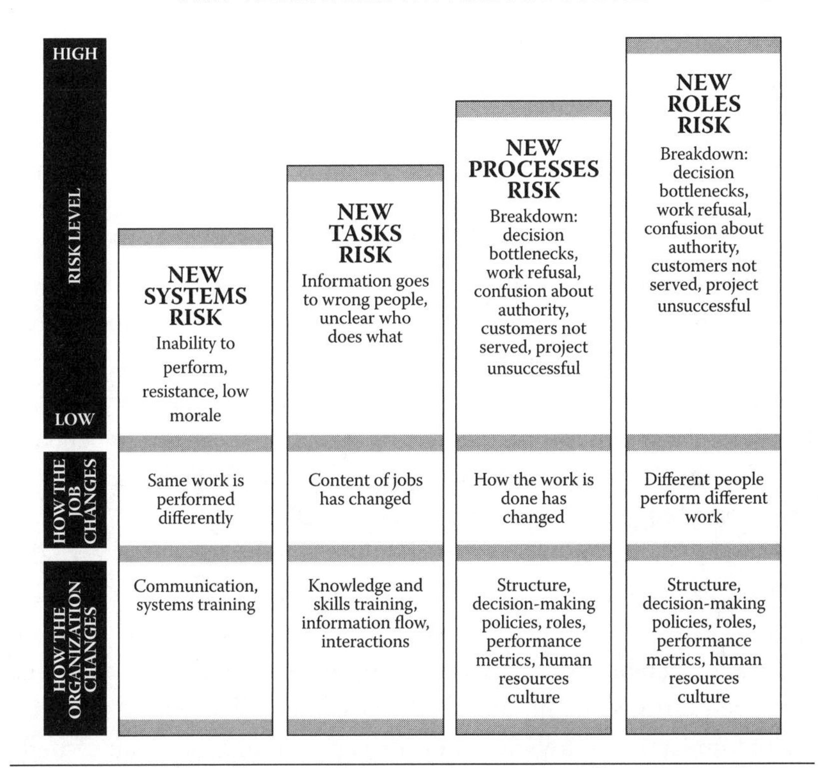

Figure 8.2 Risks associated with the introduction of new technology. (Adapted from Averett, P. The dos and don'ts of systems implementation. *Employee Benefit News 45–48*, February 2001.)

A simple Type 1 PMO does not need high-end tools, although this has been a common pitfall that companies trip into—the "more is better" line of thinking, which can be summarized as "believing that you can lead cultural and organizational change simply by installing a tool with enterprise capabilities." Too often, the result is far from positive. Figure 8.2 illustrates the risks associated with the introduction of new technology.[3] So start simply. Type 1 PMOs need the ability to do individual project scheduling and tracking. Software is required when you need more than a to-do list and a calendar; you need to manage resources and control cost.

The next level of complexity is dependencies between project tasks. Software can help identify what work can be done in parallel and what needs to be sequenced by work process. The catch is that most people sequence tasks based on logical workflow, not on resource availability. As mentioned above, this is where the simplest of project software tools can fail.

As complexity grows, resource assignment becomes crucial. Many times, in order to get resources allocated or committed to the task, the project manager has to engage in a lot of "horse trading." Frequently, the project starts moving forward before the resources are fully committed. This is a time-consuming, error-prone, and risk-exposed process in project management, one of the primary causes of project failure.

Many people are good enough at what they do that, if the project isn't too large, they can manage this without project management software, using spreadsheets, whiteboards, checklists, and the like. But, as you climb up the ladder with each level of complexity you have to deal with resource constraints that aren't obvious and dependencies that aren't obvious. And with each added level of complexity, you need a level higher of sophistication in product.

Many tools will not handle both effort and duration, but force the project manager into one paradigm or the other. Yet, your organization may need both. For example, construction tends to be duration driven—concrete takes so many hours to cure, period. At the same time, certain types of skilled labor may be needed at precise intervals on the project, so purely duration-driven planning, even on a single project under a Type 1 PMO, may fail.

The Type 2 PMO oversees departmental or divisional groups of projects. Within this type of PMO, the two main components of complexity that are added include:

1. **Interproject dependencies**. Now software needs the ability to understand what's going on in another project. At a minimum that means the ability to do some rollup and reporting across project phases. Succeeding at this begins not with the software, but with a thought process in place. How similarly are we going to manage our projects so we can compare apples to apples? What are our common phases, common terminologies? What are the actual relationships between the people doing the projects? How integrated are the projects? What's the minimal amount of standardization of procedures needed to accomplish the organizational objectives? As always, when you standardize on a process, you must weigh the benefits to the organization against the handicap to the team.

2. **Resource allocation**. Generally speaking, at this point you form a resource pool in order to track resources. If you want to do this at a departmental level, at a minimum you have to keep track of time at an activity level, not at a project level.

Most people like to allocate resources in the phase before the project is really underway. If you want to do that, you must have a role-based resource pool, one that identifies not "Bob, Mary, and Harry" but "three programmers."

Furthermore, in a large organization, or one that is geographically dispersed, where people don't know who Harry, Mary, or Bob are or what they do, you may need a role- *and* skills-based resource pool just to know who is available. An enormous amount of information regarding resources must be catalogued and entered in that system in order to make decisions about resource allocation.

Resource management, it should be noted, can refer not just to people, but to "stuff"—whatever you need to get the project done: money, equipment, people, materials. Some industries think of resources as materials, while for others it is labor, and many industries need resource management that can accommodate both.

Obviously, you need people to manage a system this complex. How many people depends on what you are going to do with it. When interproject dependencies become so numerous and subtle that the project managers can't stay on top of them, a systems administrator is required, and more than one if it is a skills-based resource pool.

At the Type 3 PMO (Strategic) level, you ratchet up the complexity because you are taking projects across departments that don't even think alike: IT, manufacturing, marketing, etc. You have to repeat the thought process described earlier, but on the corporate level.

One danger is that when people unfamiliar with the realities of managing projects get together to set up enterprise standards for projects, they will outline outrageous standards that constrain a project past the ability to be effective. Some executive-level project management training is recommended to inform executive-level staff what the minimum essentials are so that project management oversight does not become overkill.

Portfolio Management Software Issues

At the enterprise level, we encounter true portfolio management issues; selection and prioritization issues that only the newer generations of software are designed to handle. Even today, most project portfolio systems simply report on the "thought project" that has to go on at the executive level: What is our strategy? Which projects serve that strategy best? How are they doing? Where are our resources best allocated for overall corporate success?

There are decision-support tools on the market that help walk an executive team through evaluating a business case against current goals, an exercise that should filter out into committed projects. However, often, until you have some data from individual and departmental projects to roll up, you cannot make that decision because you don't know what the projects cost or where they stand.

The software requirements at this level are evolutionary because if individual projects aren't managed correctly and the consolidating or coordinating function at the departmental level is not going well, you can't evaluate projects at the enterprise level. Thus, good portfolio management is built on two things:

1. Solid data from the bottom of the pyramid
2. A structured decision-making process that ties corporate strategy to the projects, supported by that data

The capabilities typically offered by project portfolio management tools include:

- **What-if modeling**. The ability to model the effects of newly added projects and changes to available resources or project schedules.
- **Project views**. Allowing executives, project managers, end users, and workers to view the project information tailored to their information needs. The project "dashboard," which offers at-a-glance, high-level project information distilled down to the metrics needed to answer the question: Where do high-priority projects stand? allows managers to manage by exception, i.e., to take action when an agreed-upon tolerance range has been exceeded.

- **Project time reporting.** Easy-to-report status changes in tasks or projects, and the ability to record time against them.
- **Project information and status.** Task and subtask information, GANTT chart building, project searching, and the ability to drill down in a project or task from the retrieved information.
- **Resource allocation.** Resource profiling by skill, geography, business, or organizational unit, for more appropriate allocation.
- **Internet or intranet enablement.** Support for Web browsers to enable project and task review as well as the ability to change status or record time against projects or tasks.
- **E-mail notification.** E-mail notification of project deadlines, tasks, and events to individuals or groups.
- **Reporting.** Standard and custom reports should have an HTML option for publishing as static Web pages.
- **Database support.** Structured query language (SQL) databases typically are preferred.
- **Security.** Access to system features and functions based on username and password security. Other security protections also may be required, depending on the organization.[4]

A note of caution: If the organization isn't committed to engaging the "thought project" necessary to align strategy to projects, you shouldn't spend a minute or a dollar on portfolio management software. In organizations that don't perform this alignment, projects are launched without anybody in the organization doing a "sanity check" to see if they makes sense, and once launched, they are hard to kill. There are tools that can help prioritize projects, but the bottom line is that portfolio management is 80 percent thinking and 20 percent mechanics.

So, if you take a long-term perspective in looking at the project management capability in the organization, as illustrated in Figure 8.1, tools and processes need to be able to grow together. Begin building project management capability on the individual project—and individual team members—level and invest in tools that give the organization room to grow. Somewhere between Type 1 and 2, commit to a tool you can use for the long haul. As advised by researchers at the Meta Group, when standardizing on tools, organizations must first establish consistent use of the low end of the tool spectrum (and this can be pre-software, including consistent use of tools, such as templates and checklists). At the same time, they should be planning ahead for scalability of tools to a future state of project management capability, and staying abreast of emerging technology so that they can leverage developments in the field.[5]

Additional Software Concerns

Gartner, Inc. research indicates that, with the shorter duration of many IT and IS projects, knowledge professionals now move on to new projects, often in new organizational units, every two to eight months. The ability to orchestrate these workers, who are increasingly external to the enterprise, is critical. As workforces become a more diverse mix of employees, free agents, long-term consultants, and others, the difficulty of resource coordination and project administration is driving the development of new tools that support electronic collaborative practices extending beyond the enterprise. These tools offer features that extend project management to support project accounting, online analytical processing, management dashboards, Web schedule publishing, Web collaboration, project document support (knowledge management), estimating, and much more.

Collaborative Tools: Promise and Frustration There is so much good technology for collaborating and sharing information that it is tempting to focus on the functionality of products. But the real challenge is to design the social side of information technology. And, there is only one rule for good "social design" of collaborative technology, according to community of practice guru Richard McDermott: *Make it easy to connect, contribute to, and access the community.*

Software should make it easy for community members to connect with each other, and to use information from the community's knowledge base. This is more a matter of how seamlessly the software integrates with people's daily work patterns, with the kind of knowledge they need to share, and with the way they envision their community's domain and navigate within it, than with specific technical features of the software. For example, when saving documents for the team or posting for the community involve the same number of steps in a familiar software product, the friction involved in collaboration is reduced. *Friction* is a useful way to think about barriers to communication or collaboration within a community. How difficult is it to connect, contribute, or find help? Does it require going into a new program, accessing a separate area of a network, learning new computer skills? Is the collaborative software intuitive? The more special effort it takes to connect, the more friction is created. The greater the friction, the less likely people will take the time to connect.

This is why face-to-face communities have an edge over global ones. If you run into members of the community at lunch, the friction involved in sharing is minimal. The community "space"—Web sites, databases, libraries—should be organized according to some principles or taxonomy, preferably one that reflects the natural way community members think about their field or topic. McDermott gives the example of a community of geologists who preferred their knowledge bank to be visualized like a map, a kind of document they used in their work and were comfortable with. A well-designed knowledge community offers a bank of information that is easily navigated; users can intuitively find familiar "landmarks," use standard ways to get to key information, and browse for related items.[6]

There are a number of tools commonly thought of as "knowledge management" software:

- Intranets
- Document management systems
- Information retrieval engines
- Relational and object databases
- Electronic publishing systems
- Groupware and workflow systems
- Push technologies and agents
- Help desk applications
- Brainstorming applications
- Data warehousing and data mining tools

The problem with this list is that it addresses *explicit knowledge*, but ignores the human repositories of *tacit knowledge*. Such approaches to knowledge management display the technologist's fantasy that a system will do all of the work. But, in project management and knowledge management, as KM writer and consultant Wally Bock has written, "Thinking at all stages is mandatory."[7] While the above software tools can help to save and organize information, the social design of a PMO/project management community will be, ultimately, more crucial to success than the purchase of the most sophisticated tools.

Increasingly, PMOs are coming to realize that social media (wikis, social networking sites such as Linked In and Facebook, blogs, Twitter, YouTube, and the like) can all be used to keep project teams in touch with each other, build competency by making it easy for experts to share their knowledge, and for novices to find information, and add sheer fun and ease-of-use to the daily workload. In addition—despite some security concerns—these tools allow project managers and their teams to inhabit a community of practice that extends beyond the company, and draws on the best that the "project management universe" has to offer. The laptop is the new water cooler, but the crowd around it numbers in the thousands, and their tacit knowledge is searchable. It's really kind of mind-boggling. These tools have met with resistance in some organizations, but, in others, they have been accepted and shown to maximize productivity. The CIO of the Kennedy Center notes, "Where traditional project management focused on managing the inputs and outputs of one project, with the goal of scheduling, we now focus on managing the processes and work of a project, where the real goal is successful collaboration. Social media enables this collaboration."[8]

What Are the Best Practices for Selection?

Most new software products claim to support PMO functions. Of course, just because it doesn't say "PMO" on the box, doesn't mean it isn't useful to the PMO. Most of the enterprise suites' tools have jumped on the bandwagon offering enterprise project management and portfolio management features. For more information and, in many cases, demos for each product, visit the company Web site. Extensive linked lists of project management software tools are available at www.pmforum.org, www.allpm.com, www.projectconnections.com, and other project management portals and organization sites.

Make selection a systematic and carefully thought-through project. The most experienced software selectors recommend at least three stages of analysis when selecting software:

1. **User analysis**. How skilled are the personnel who will be using it? A third-party, objective assessment is best to determine this.
2. **Needs analysis**. What are they going to do with it? What kind of reports are needed? What needs to be tracked? What problem are you trying to solve? Be specific.
3. **Vendor analysis**. Are they financially stable, with a good support track record with other clients?[9]

One of the best processes for software selection we've seen was offered in the preface to a PMI survey of project management software products. It was compiled of the following eight steps:[10]

1. **State the business problem driving the selection project**. Does the infrastructure need upgrading? Do needs exceed the capacity of the current system? Or is project performance below expectations? The first two problems are more easily addressed: document what data is not available, how it will be used when it *is* available, and how that information will help meet strategic objectives. Define the information required in detail in order to generate requirements for the software. The third problem is more intractable. If the process is flawed, new software won't fix it. Analyze the reasons behind poor project performance before launching the selection project.
2. **State improvement objectives early in order to define requirements in terms of the features that will actually be used**. This means documenting the current state of things, quantifying the improvements, and finally, defining the information required to support the improvements. Document what information is not available from current system. Treat this as a visioning exercise, using subject matter experts or consultants as appropriate. State the high-level goals of the project management environment, such as *reduce time to market of new products by 10 percent*, or *reduce project effort by 8 percent in 18 months*. Why can't these goals be met with the current system? Document how the software should help meet these goals.

3. **Document or reengineer the project management process**. If the current process is inappropriate for solving the business problem, meeting the improvement objectives, and meeting the goals of the project management environment, document a new process before continuing with software evaluation.
4. **Map the project management process to the appropriate category of software**. This is the most labor intensive step. Schedule a series of detailed sessions to walk through each step in the process. Make detailed notes describing the inputs, outputs, and processing required to support the process. Involve subject matter experts, management, the current system, and IT staff. Produce a detailed document explaining how the system will support the process for each step. Capture the data elements required. There are many ways to achieve an information objective, so identify *what* is required, not *how* the software should make it happen.
5. **Generate a requirements list**. Most critical requirements will be uncovered during the mapping process. Articulate these and the short list will identify the best possible matches to the requirements.
6. **Select the short list**. A short list of two products is ideal; more products shows that requirement definition has not surfaced a differentiating feature, and almost any product will do. A detailed set of requirements based on a sound process should always result in identification of a need that can only be met by one or two vendors.
7. **Test products yourself**. Invite vendors to introduce you to their software. Testing them with data that actually represents the business problem facing you will demonstrate which products handle the requirements in the way you expect. Use experienced people to evaluate the results. They have an understanding of what results are expected, can articulate the pros and cons of the differences between how the packages perform, and can comprehend the potential benefits of differentiators between products.
8. **Decide/judge the vendors**. Software evaluation efforts can take as little as two days or as long as four months. There is no one best product; seek not "the best," but one that meets the

> improvement objectives of your investment. One important element in selection should be careful consideration of the vendor's strength and weaknesses. You should assess their reputation (years in business, customer base, financial condition, proven track record, alliances with other vendors, long-term strategy). Judge how they deliver software and support. Do they use partners and service providers? Who will deliver and implement the software? Do they have a local presence? Are there performance guarantees? Is there adequate training and documentation? Finally, what technology decisions have they made and how do those impact their future plans? If, for example, you choose a vendor that has "hung his hat" on a task-management approach to project management, as discussed above, will you be limiting your company's future ability to evolve into a more sophisticated project management practice?[11]

To this, we might add that at a certain point, there is a diminishing return on the time invested in evaluating software. *Computerworld* has pointed out that most of the leading software offerings will meet most business requirements, and fewer than 50 percent of the planned features actually end up in implemented systems, regardless of the upfront analysis. The most successful software selection efforts, they argue, strive for speed and results by setting a firm schedule, avoiding requests for proposals, narrowing the field quickly, and getting all the costs on the table.[12]

Finally, some of the most common pitfalls include focusing too much on price and getting locked in to a product that doesn't really meet your company's needs; not understanding the knowledge level of the staff and presenting them with a product that is either far too difficult or insultingly basic; failing to get senior management buy-in so that software is investigated, but never purchased, or installed but ignored by users; buying software that doesn't easily integrate with existing architecture; and thinking the tool is the solution, when, in fact, training, communication, process improvement, and organizational culture drive project management success.[13] Don't fall into the trap of thinking that software is the keystone of a project management improvement initiative: People do projects, the tools are secondary.

Rollout: Putting the Tools to Work

Now that you have made the decision, how should you roll out the software? Everyone will want to touch, feel, and use the new software immediately. It's like a new toy and sometimes it is treated that way. That's okay as long as you use logic, reason, and patience in putting it together and introducing it to the organization. To do this, we suggest the following:

1 Proceed slowly and incrementally. Don't promise too much too soon. Phase in the capabilities of the tool in a building-block approach. Perhaps work on basic scheduling elements first, then resource management, and then full-cost estimating. Phase in the tool *throughout* the organization. Don't roll it out to everyone all at once. Before you do anything, think through your project management requirements and build a logic structure in the database that supports your business needs. Don't curtail this part of the effort … the more time you spend thinking through the database configuration, the happier you will be in the long run.
2. Pilot test using select projects and portions of the organization to show successes and get supporters. This almost speaks for itself. The benefits of pilot testing are more than "working through the bugs," they give you the opportunity to have successes (which you can brag about) and earn supporters throughout the company. It is always helpful to have strong advocates that openly support and willingly talk (positively) about the new software tool.
3. If needed, fix the process first and teach your people the new principles, then adapt the software to the new business rules. There is an added complication if you plan to introduce new processes along with a new tool. If you can, it is best to keep the process somewhat similar, and introduce the new tool using the existing process. In this situation, folks will learn and apply one new element versus two. In some cases, it is entirely unavoidable and you have to introduce new processes and a new tool concurrently, e.g., perhaps your current method doesn't work or a process doesn't exist. In such situations, make sure your people understand and accept the new

principles of the process before you expose them to the new software environment that will make it efficient.

4. Keep the software environment and database structure simple, especially at first; don't get too detailed or complex. I've learned this one first hand. In one job I managed, we were overexuberant and built a new system that met everyone's wishes and did everything imaginable. It took awhile to deliver and, once it was delivered, it was "too much." Keep it simple. Keep it straightforward. Stay basic and you will be successful. Don't put too many details in the core structure of the database. If you need more, add the expanded capabilities in future increments.
5. Implement software components to focus on the most pressing needs of the organization, e.g., planning and time entry before cost estimating. This ties back to one of the earlier comments of proceeding slowly and incrementally. Your incremental implementation should start by addressing the most pressing needs of the organization. This may be project scheduling or resource tracking. In the implementation, the best thing to do up front is to take care of the organization's most important needs. Identify what those are and structure a rollout plan accordingly.
6. Train, train, train, train and mentor, mentor, mentor, mentor. People will make it happen, but you need to make sure they understand what to do and how to do it. Conduct multiple, short training sessions. Make sure you emphasize in the training sessions why the organization is implementing the tool (and process). Don't assume people know how to use the tool or that they understand the new process. Offer training using multiple avenues: classroom, intranet, handouts, one-on-one, handbooks. Focus heavily on training. You need to make sure everyone understands the basics; don't assume the word is relayed through management. Show your continued support after training by having a formal or informal mentoring program to help when they encounter specific problems afterward.
7. Plan the implementation and communicate the objectives to everyone; do it again, again, and again. Make it very clear why you are implementing the new software and explain the

benefit to everyone. This should be done way up front and accomplished over and over again. Plan the implementation so the objectives are clearly laid out. If you are planning the implementation in phases, make sure there are clear objectives, benefits, success criteria, and identified risks for each phase. Whenever there is change in an organization, people respond better if you can explain the benefit and outline specific accomplishments. Then they understand what to expect.

8. Get senior-level sponsorship and support; you need more than just the words, you need expectations. In other words, the senior staff won't help you with simple lip service. They need to believe in the reasons behind the new software implementation and tell you what they expect. The senior staff needs to personally tell the employees in the organization their expectations and that they stand behind the new implementation. Obviously, it is up to you to make sure the senior staff understands the whats, whys, whens, and what-fors. Saying it once won't be enough, you will need to repeat the same reasons every chance you get.

In summary, software tools are a good complement and serve as a good foundation for PMOs. Relatively speaking, the hard part starts after tool selection as you roll out the new tool environment within the company. Make sure you plan the implementation, attain senior-level support, communicate the plan, and proceed at a reasonable pace with reasonable expectations.[14]

Notes

1. James S. Pennypacker, *Resource Management Challenges: A Benchmark of Current Business Practices* (Glen Mills, PA: PM Solutions' Center for Business Practices, 2008).
2. M. Light and T. Berg, "The Project Office: Teams, Processes and Tools," *Strategic Analysis Report* (Stamford, CT: Gartner, Inc., August 2000).
3. Patricia Averett, "The Do's and Don'ts of Systems Implementation," *Employee Benefit News*, February 2001.
4. Light and Berg, "The Project Office."
5. M. Light and T. Berg, "Application Delivery Strategies," *1999 Trend Teleconference Transcript* (Stamford, CT: META Group, 1999): www.metagroup.com.

6. Richard Mcdermott, "Knowing in Community: Ten Critical Success Factors in Building Communities of Practice," *Knowledge Management Review* (May/June 2000).
7. Wally Bock, "Knowledge Management 101," *Intranet Journal*, http://idm..internet.com/articles (article originally appeared in *Briefing Memo* at www.bockinfo.com).
8. Alan Levine, "Social Media Tools: An Introduction to their Role in Project Management," in *AMA Handbook of Project Management*, 3rd ed., eds. Paul Dinsmore and Jeannette Cabanis-Brewin (New York, NY: AMACOM Books, 2010).
9. Max Feierstein (LDS Group), interviewed in *PM Network,* September 1996.
10. *Project Management Software Survey* (Newtown Square, PA: PMI, 1999).
11. David Golan, *Call Center Solutions* (August 1999).
12. Michael W. McLaughlin, *Computerworld* (June 19, 1995).
13. Feierstein, interview.
14. Dianne Bridges, "Software to Support the Project Office," *Project Management Best Practices Report* (March 2000).

9

Changing Organizational Culture

Corporate culture—the beliefs, behaviors, and assumptions shared by individuals within an organization—includes such things as procedures, values, and unspoken norms. Culture can have a significant influence on how well strategy is executed in organizations. The importance of achieving strategic objectives, how performance is communicated, whether or not changes create competition or cooperation, who can access and use technology, whether or not decision making is done in command-and-control environments or by self-directed teams, how functional units work with each other—these are just a few of the issues of culture that need to be addressed in creating a structured approach to executing strategy via a Strategic PMO. Some best practices, identified in our *Strategy & Projects* research study, include:

- Project management is valued throughout the organization.
- Risk planning is an important part of the organization's culture.
- A focus on strategy execution is an important part of the organization's culture.
- Senior management is trusted, and consistently rewards successful project behaviors.
- There is a shared understanding and commitment concerning the organization's long-term objectives and its strategy for achieving them.
- The most often used practice by high-performing organizations is having leadership that is trusted. High-performing organizations are significantly better than average at having senior management that consistently rewards successful project behaviors.[1]

Needless to say, these cultural characteristics may seem out of reach for the organization that is operating in an ad hoc mode with little in the way of existing project management structure, process, or culture. And, although the culture of an organization can be difficult to alter, it is not impossible when the change is approached as a project in itself.

Climate or Culture—Which Do I Change?

What's the difference between climate and culture? Culture has already been defined as a set of shared beliefs, values, and expectations. A climate, on the other hand, consists of the tangible things that make up a culture, such as policies, procedures, habits, and routines—all the things that define how things are done in the organization. For example, the compensation structure of the organization (how people are rewarded) is a matter of climate. If quantity is rewarded over quality, the practice will be to crank out as much work as possible without regard to errors. On the other hand, if people are rewarded for producing a quality product every time, and even have the power to shut down an assembly line when defects are detected, the practice will be to produce quality. This practice will lead to a change in culture, one where employees believe what the organization's leadership believes in and values: quality.

A project example illustrates the point further. Let's say the company project management methodology says to create a risk management plan—an issue of climate—but management never asks to see it. The message is clear: Don't waste your time doing risk plans. On the other hand, if management requires the risk plan to be updated and presented at every project status meeting, the message is clear: Management values and rewards risk planning. This change in the way things are done, a change in climate, leads to a shared value, that risk planning is an important part of the organization's culture. In fact, years later, when someone asks why risk planning is done so religiously, one might give the response that: "It's just the way we do things here," not knowing the origin or evolution of the practice that led to the climate, which changed the culture.

Change to the culture will not occur through declarations, new mission statements, or a big party, unless these proclamations are backed up by changes in the everyday practices and procedures that define what people should do, how they are evaluated, and how they are rewarded. These changes in the way things are done will lead to a new belief that the new way is the right way, a matter of culture. We demonstrate from our experience in the project management field that improvements in the way projects are managed leads to bottom-line success. Another way to think about it is to "act as if" the desired belief system were already in place to bring about the desired change. After improving project management practices and providing members a chance to see that improved practices lead to success, individual beliefs in the merit of improved project management process will follow. Before long the organization will naturally assume that the most effective and efficient way to manage complex undertakings is to organize and execute them as projects.

Organizational Needs: Flexibility and Creativity

Companies today need employees who can think fast on their feet, innovate, take responsibility, and "go with the flow" of change. Yet, although these are the characteristics companies are seeking, too often they are the characteristics least rewarded.

Companies frequently speak of vision, but express their vision as a concrete set of goals and objectives, solidified into rules. A vision must be the inherent way we function, the essence of what we are, the way we feel about ourselves as a unit and as individuals. If we know only what the goals and objectives are, anything new that does not match our expectations throws us into confusion and chaos. It is the difference between the internal and the external. We must always be able to evaluate situations and opportunities according to the core spiritual and emotional assumptions of the organization.

Goals and objectives can create tunnel vision. As the organization transitions from a functional to a project culture, we must constantly reevaluate and reconsider.

- Do our goals focus efforts so much that we are unable to see opportunities?
- Do they keep us from making the paradigm shifts necessary to anticipate the needs of customers or make it possible for employees to soar?
- Do they lock us into only the assumptions that were available when we set them?[2]

Creating a Project Culture: From the Top

We have established that the culture cannot be changed unless the practices that make up the organizational climate are changed and rewarded. In the project management context, this entails establishing a whole set of new behaviors, starting with creation of a project management methodology, defining what is required, when it is required, and how to do it. A complete set of instructions, forms, templates, and tools is necessary to ensure consistent, repeatable performance across the organization. Next, a training program, tailored to the new methodology, but grounded in the project management body of knowledge, is necessary to teach and reinforce use of the methodology. Finally, management must *require* consistent application of the methodology and reward successful project behaviors. This is usually thought of as bringing in a project on time, under budget, according to specifications, with a satisfied customer, but also can be in knowing when to kill a bad project.

Specific Guidance on Changing Your Culture

Schneider, Brief, and Guzzo have published an excellent paper entitled "Creating a Climate and Culture for Sustainable Organizational Change" that lists six steps to implementing what they refer to as "total organizational change (TOC)."[3] In it, the authors suggest following a six-step process to introduce any TOC into an organization. We've added an initial step to make it a seven-step process, discussed below.

1. **Assess the need for change**. Do you even *need* to change the organizational culture? Chances are if you are reading this book and considering implementing project management across the organization, the answer is yes. But, to be sure, perform an assessment to determine your level of project management maturity. If you are at Level 1 of the Project Management Maturity Model, just beginning to define processes and practices, you should read this chapter carefully before "beating your head against the wall" trying to get everyone to embrace a new way of thinking.
2. **Ensure the organization is prepared to handle a major organizational change**. By "prepared" we mean the level

that people in the organization believe that the change is necessary and that management will support the change. And beyond that, we also mean the actual level of commitment senior management will give to the change. Schneider et al. phrase their discussion in terms of trust.

If the organization's management is not trusted, any attempt to change will be treated with skepticism on the part of the organization's members. We have all heard statements like: "This is just the flavor of the day." "This too shall pass." "This will last about a year then we'll be into something else." "This is just like TQM—here today, gone tomorrow." These are statements that reflect distrust in the organization's leadership. To put it bluntly, change will be difficult in an organization where the walls are papered with Dilbert cartoons. On the other hand, such a place probably needs the changes discussed in this book more than most. Ask the following questions before attempting to install a Strategic PMO:

a. Is employee morale high? High morale is usually associated with a high level of excitement and commitment about belonging to the organization.
b. Does the senior leadership have a history of successfully implementing major changes? If so, people will naturally assume that the pending change to a project management approach has been well thought out and that senior management will see the change through.
c. Is management known for tackling tough decisions and doing the "right thing?" If so, management will be respected for its integrity and perseverance.

On the other hand, if the answer to these questions is no, any attempt to implement the sweeping changes necessary to implement a Strategic PMO will be met with skepticism. It is advisable not to attempt the change until trust is established—something that will not happen overnight.

3. **Is the proposed change consistent with existing organizational climate and culture?** Major changes will have a greater probability of success if they are undertaken in incremental steps that are consistent with the existing organizational

culture. Change that is consistent with the existing culture will be seen as user friendly, nonthreatening, and will be more likely to be accepted than change that is radical in nature. For example, if your organization is structured as a vertical, functional bureaucracy, where all the power is vested in functional management (a traditional functional organization structure), then changing to a cross-functional, lean, project environment in one major step will be difficult and time consuming. Implementing a Strategic PMO will cause quite a stir, as functional managers are asked to share power with a "new kid on the block." Radically changing the culture will require a significant amount of time, effort, and attention in such an organization.

On the other hand, if your organization is already managing projects, issuing project charters, and empowering project managers to reach laterally across the organization for resources and support, then establishing a project office will be seen as the next logical step to improving the practice of project management. Even if you have experienced project failures, often a strong move by management to do something about it is seen as a positive step.

In either of these cases, it is inadvisable to go from no project management processes to a Strategic PMO in one step. In deciding how ambitious to get in changing the culture, consider the following four dimensions of organizational climate:

a. The nature of relationships. Are they contentious? Is conflict the order of the day? Or, do people work collaboratively toward solutions, freely sharing information?
b. The nature of existing hierarchies. Is the organization rigidly structured or is the structure flexible and agile?
c. The nature of the work itself. Is most of the organization's work project or process oriented? Does the work have a defined start and end? Can specific objectives be established? Are resources from more than one department involved? If so, the work can be classified as project work. Otherwise, it is continuous in nature and not suitable for the project management approach.

d. The focus of support and rewards. Are people rewarded for achieving results or just showing up and putting in their time? This question addresses one of the fundamental differences between bureaucratic and entrepreneurial organizations. Project management rewards those who can bring projects in on time, within budget, according to customer expectations.

The proposed change to a project culture should be examined, and the impact on each of these dimensions analyzed, to gain an appreciation for the effort that will be required to bring about the change.

4. **Plan the change in as much detail as possible**. Nature abhors a vacuum. If information about the change—what is happening and why—is not continuously shared with the rest of the organization, the rumor mill will defeat you. For example, in working with one of our clients engaged in establishing a project office in its IT division, we found in a joint meeting of senior management that the project office implementation team had not been keeping members of the CIO team (a group of senior IT managers) informed of the status of the implementation effort. As a result, they assumed the effort was having difficulty or had even been canceled. Without their support, there was literally no chance of success. Following the meeting, the implementation project manager made it a point to personally keep all key managers informed.

Following some commonsense guidelines will help you with this step:

a. Formulate and widely communicate the plan for the change you are considering.
b. Specify, in writing, the objectives of the change. Use the same guidelines used in formulating project objectives; they should be SMART: Specific, Measurable, Agreed-to by stakeholders, Reasonable, and Time constrained.
c. Follow up dissemination of the objectives with written systems, methodologies, and procedures for the change. Let's say you are trying to publish a new project management methodology as part of your effort to change the

organizational culture. You must first ensure that you establish a cross-functional team to help you write and review the methodology. You must also make sure the team members keep their managers informed of progress; make sure they are excited, onboard, and feel a sense of ownership for the new guidelines. And make a special effort to apprise management at every opportunity. One final thing: Find out how to get information out directly to everyone in the organization affected by the change.

d. Follow up written methodologies and procedures with support systems and rewards for following the new practices. The support will come from the new PMO. In a Type 3 Strategic PMO, the project manager reward system will be driven by the PMO with significant input and influence from functional management. Team member performance will primarily be determined by functional management with input and influence from project management. Work closely with functional management and Human Resources (HR) to structure rewards (and they don't all have to be monetary) for those who follow the new practices and who achieve better results.

In short, specify why the change is necessary, what is threatening the current organization, and how the proposed change will defeat the threat. Spend time and money developing the methodology, processes, and policies. Distribute the methodology and other literature widely. Make it clear to people that they will receive training, will be expected to implement the new practices, and will be rewarded for doing so. Going half way with this step will lead to disaster.

5. **Ensure the reward system is structured to motivate employees to focus on implementing the project management methodology**. People are smart. They figure out what the organization rewards and they focus their energy on doing those things well. Almost everything else is ignored. Management must determine specifically what behaviors will be rewarded in advance and tie appropriate behavior to both

financial and nonfinancial rewards. Project management is a well-defined set of processes and activities. These processes are published and distributed around the world by the Project Management Institute as *A Guide to the Project Management Body of Knowledge*.[4] In changing to a project culture, these practices, articulated in a project management methodology document, will be required. We've already discussed how, by implementing these practices and achieving success, the organizational culture will be changed over time. But, unless management insists on, and rewards, behavior associated with the consistent application of the methodology, this approach will fail, and all the time, effort, and money devoted to implementing the change will have been wasted. More importantly, the organization will not achieve project success, which will probably spell disaster in a very competitive marketplace.

6. **Allocate resources to maintenance of the new system**. To ensure the change to a project culture is sustained, resources must be allocated to maintenance of the new culture. The maintenance and support of this new approach comes from top management, and is operationalized through the project office. The PMO, as we have seen, supports the practice of project management and, therefore, the culture, by becoming a center of excellence, by providing organization-wide support in a variety of ways, by owning the project management process and publishing the methodology, by providing a home office for project managers, by providing mentors to projects throughout their life cycles, and by training project managers, team members, and other managers.
7. **Monitor the progress and effectiveness of the change to the organization and adjust as necessary**. This step is fundamental project management practice. For example, in developing risk mitigation strategies, the project team develops the strategies, implements those that become necessary, checks to see if the strategies are effective in reducing risk, and adjusts the strategies to ensure they are effective. In a similar manner, when management is implementing a change to a project culture, it

Pilot Phase	Deployment	Functional	Integrated	World Class
Definition Responsibilities Training Process **Champion identified**	**Consistency** Tools PM role **Project manager qualification** Selection Training	**Consistency** PM process Data Tools **Project manager capabilities** Broad authority Certification Mentoring	**PM structure** **PM tools widely used** **Project managers certified** **Performance Improvement**	**PM capability used as benchmark by others** **PM directory contributes to best-in-class performance**

Monitor progress and adjust as necessary

Figure 9.1 Assessing effectiveness of change.

must check periodically to ensure the desired behavior is, in fact, occurring. If not, either changes to the practices must be made or more emphasis put on changing and rewarding the behaviors. Either way, performance must be monitored and variance minimized to bring about lasting change. To monitor progress, the project office might consider a simple graphic, such as that shown in Figure 9.1, to guide its long-term effort. It might also be advisable to periodically reassess project management maturity to measure real improvements in project management practices, artifacts, and results.

Measuring for Results

Daniel Tobin, in his book *Re-Educating the Corporation*,[5] identifies the need to create relevance, value, quality, and measurement criteria in order to transition successfully to a new corporate culture.

Relevance If the employees do not find that the proposed change is relevant to their current working environment, they will not support it. To make a change relevant, the context in which that change is perceived must be changed. For example, Tobin suggests that functional and project managers performance measurements to reflect the value now placed on teamwork.

Value Employees need to understand the value that the change has to their work and their design performance measurements. Tobin cites the example of departmental employees who, if they are to adopt team-related work methods, must view the teamwork as a way of making their jobs easier or making them more effective in their work. Unless they perceive the value added by teamwork, no amount of training will persuade them to work as a team.

Measuring A functional-to-project organization transition should be measured by results that are based on a definition of quantifiable objectives. As such, the entire program needs to be measured in its entirety; the individual pieces should not be evaluated as an end onto themselves, but rather in how they support the overall transition effort.

Change Processes

Typical approaches to dealing with change and transition in an organization are to attempt to predict the future and manage to that potential reality, or reactionary management, waiting for situations to occur and responding accordingly. Neither is likely to succeed. The following approaches can lead to a more long-reaching and effective change and transition implementation.

The Scenario Approach

We can't predict the future, yet we all spend a great deal of time attempting to do just that. It is much more productive to question what potential scenarios might occur, what strategies can be used to deal most effectively with each of them, what resources will be needed, what skills will be required, and what critical competencies do we have right now that we can build upon.

The trouble with planning based on predicting the future is that it implies we can control the future, creating an illusion of certainty. "I've thought it out, made my plans, therefore, this is the way it will happen." From these illusions specific objectives are established and from that point on, any contradictory evidence is ignored. These internal contracts too often form an impenetrable box that isolates us from reality. It is like the old saying: "Don't confuse me with the evidence. My mind's made up."

The perception of order can create an artificial sense of stability; in reality, it is the most unstable environment of all for it lulls people to a passive state where they are least of all prepared to deal with the unexpected, which surely will occur.

The scenario approach, on the other hand, is based on plural possibilities for the future. It encourages questioning; it recognizes the reality of ambiguity and expects change. It creates a "memory of the future" that prepares people to deal effectively and confidently with whatever occurs, and makes it much more likely to direct the course of events in the most positive direction. It also teaches people to be effective problem solvers and it reduces fear of the unknown, which are critical attributes for individuals and companies in the twenty-first century.

Scenario planning demands that everyone look outside their own immediate interests into the broader world outside: social values, technology, economics, environmental issues, political thinking, changing cultures, etc. It encourages relationship awareness and network building, the foundation for synergy. It is the essence of project organizations that forms a successful partnership between their internal managers and their project managers.

A Project Management Approach to Change

Any change that is significant enough to engender transition within the company is significant enough to be considered a project. For example, going from a functional to a project organization will require:

- A defined scope (the magnitude of how business operations will be modified).
- Associated cost parameters (are new systems required, are new personnel required?).
- Risk issues (opportunity risk as well as the risk of not being competitive).
- Quality issues (ensuring that the same quality of service/product is offered throughout the transition).
- Designated time frame constraints (will this occur over one month, six months, or one year?).

- People who will be involved and need to be managed (how will roles and responsibilities be defined, will this require reorganization within the company, will new skills need to be learned?).
- Major communications issues (how do we communicate with our existing customers, our employees, our stakeholders?).

The problem with most corporate change and transition is that it is poorly managed. Some changes are treated like a project (business process reengineering is a notable example), but most are not. The attitude of management tends to be one of "survival of the fittest," and then they wonder why productivity sinks. Employees need to understand the value and rationale behind corporate change and transition. They need to understand how, when, where, and why this is occurring. Project management is an ideal way to manage all corporate change and transition by defining what will occur, when it will occur, how it will occur, and where it will occur. By developing respect on all sides (employee, customer, and corporation), the organization puts itself in a position for successful change and transition.

How does an organization move forward with a business or personnel change in the most productive manner?

- Begin with defining the project parameters of the change.
- Develop a project plan, elicit comment and feedback, and get buy-in from those who will be affected.
- Baseline the project plan and manage the customer expectations of the end deliverable.

Readying the Troops for Battle

Besides developing a solid project plan and managing it, we can prepare the people for cultural change by helping them:

- **Accept Ambiguity.** *"Let's wait until we know for sure before we make a decision"* were no doubt the last words of the dinosaurs. The need for certainty must be replaced with not just an acceptance, but a firm belief in the power to evolve in a positive way with whatever the future holds.

- **Prepare for Possible Scenarios.** The reality is that we constantly create possible scenarios for the future. David Ingvar, the head of the Neurology Department at the University of Lund, Sweden, calls this the "memory of the future." The human brain automatically attempts to make sense of the future by testing possible plans of action; therefore, when the situation actually occurs, there is a "memory" of how the choices were sorted before it occurred.
- **Have Fun in Order to Survive.** Creativity, imagination, insight, and intuition are most likely to occur when we're having fun, when we're not taking ourselves or our activities too seriously, when we feel free to react naturally, to just be.
- **Forget Consensus—Conquer through Collaboration.** Consensus guarantees mediocrity because it focuses on reaching agreement rather than emphasizing deeply held convictions, which most often exist because people have expertise or knowledge in an area. Collaboration focuses on people's strengths and creates synergy.
- **Adapt to Life in the Chasm.** The gap between the known and the proposed, between our knowledge of the past and our image of the future, can seem like a huge chasm. How does one learn to live with optimism and faith in oneself and in the organization's ability to make a safe bridge across? First, by acknowledging the reality of the gap. Second, by looking for congruence between the sides, most often based on values. Third, by being excited by the possibilities.
- **A Task Is for Today—A System Is for Always.** The workplace is shifting from one of obedience and domination, to cooperation, mutual respect, and shared responsibility. Networks of collaborative relationships are creating transformational learning based on integrated systems of people working together. In a rapidly changing world, it's not enough to learn a skill, we must create structures and processes in order to reach higher levels of performance. We must have a system, rather than a task perspective. It's big picture thinking. It involves asking such obvious questions as: "Are we going about this in an effective way?" "We seem to be jumping all over the map

in this discussion. Let's take a moment to focus on identifying the problem." "What is the scope of our project?"

- **Rules for Successful Culture Implementation.**[6] The success factors examined here can be summarized in a set of rules for organizational culture change. The more rules that are followed, the greater the chances for successful implementation.
    - Become a learning organization. Embrace new ideas, new concepts, new techniques, and make them available to everybody.
    - Establish clear communications processes and media.
    - Record and praise accomplishments and heroes who support and demonstrate the concepts required in the new culture.
    - Establish a flexible, central structure that provides a critical core for all implementation efforts.
    - Accept risk and proceed judiciously. Strive to extend the culture throughout the organization despite the inherent risk of change.
    - Know and publish boundaries for the culture. Ensure a common understanding of what the culture is intended to be and what it isn't.
    - Evaluate and prove the economic value of the culture.
    - Involve everyone.

Overcoming Barriers to Change

In her excellent paper on change management in the project organization,[7] C. J. Walker discusses how entrenched divisions of power in a functional organization poses a barrier to change.

People and Power

One of the major obstacles in transitioning from a functional to a project organization is the perception that functional managers must give up their power base. To become a project organization means that project managers become, in effect, temporary managers of resources. Their decisions directly affect the organization as a whole and the functional managers become leaders of resource pools with the responsibility for developing and managing human skill sets that then will be "on loan" to projects.

To ease this transition, create fluid roles and responsibilities that integrate functional managers into the total organization. Their focus in the past has largely been reactive, identifying and responding to internal drivers and fulfilling current business needs. Today's companies require functional managers who are capable of focusing on the big picture, proactively anticipating future needs, and actively participating in multiple areas and projects at the same time. As projects become the organization's driving force, project managers must negotiate with functional managers to acquire skilled and flexible resources to fill specific short-term project needs. In addition, the organization must allow the functional managers the latitude to staff appropriately to meet both the ongoing operational needs and the project needs of the organization, and there must be an acknowledgment and reward system built into the process that recognizes the value and importance of supporting project needs. The functional managers and the project managers must work together as a management team, developing a work force that is flexible and mobile.

The second step is recognizing the impact to the individual when the organization changes from a *functional* focus to a *project* focus. The basic premise of project management is that a clearly defined venture with known objectives and goals will be completed in a finite period of time. Once the starting gun goes off, the project manager is expected to lead the way down a direct path to the goal line. This can create a major destabilizing force, distracting people from their on-going priorities and causing resentment throughout the organization.

The third step is understanding the evolutionary nature of projects. New evidence will emerge throughout the life of the project, changing initial assumptions and requiring paradigm shifts that substantially change the comfort level of employees.

The project manager must be sensitive, not only to directing the activities of the project, but to gaining the support of the people. Futurist Joel Barker describes three stages in the life of a paradigm and defines the people most likely to be comfortable in each stage.

- *Paradigm Shifters* realize the current paradigm does not solve all the problems; they determine the issues and create the new paradigm.

- *Paradigm Pioneers* lead the implementation of the change; they establish the parameters and rules and provide the impetus to move the organization forward.
- *Paradigm Settlers* institutionalize the change; they root it, refine it, and ensure excellence in operating under the new system.

A project responds to a defined need, lifts the element up out of the on-going work of the organization, solves it, and reinserts it back into the organization. The organization is changed by this process. Invariably, the paradigm has Shifted, the Pioneers have left and too often the Settlers are asked to proceed without having been involved in the change. They are expected to integrate the change into on-going systems, and to create excellence without having the opportunity to develop the required skills.

The project cannot be isolated from the functional work of the organization. Therefore, the project manager must be integrated into the operation, serving as the transition guide, working directly with functional managers to identify those employees who will be most comfortable and, thus, most productive in each stage, pulling the Settlers into the project early enough for them to identify and develop the skills they will need later in order to ensure excellence and their willingness to accept the new paradigm.

The fourth step is understanding the impact that organizational culture has on project success. In a January 1995, *PM Network* article, Dan H. Cooprider said, "Culture is the key ingredient of a company's success. Discovering, defining, and changing that culture is the business challenge of the [19]90s. … As more attention is placed on corporate culture and more projects are undertaken to change or manipulate that culture, project managers will have a corresponding need to understand and manage aspects of culture pertinent in their project."[8]

In the meantime, those working to integrate the SPMO will be faced with that old organizational demon: politics.

Organizational Politics, Roles, and Responsibilities

The Strategic PMO must manage the politics of being a new player at the corporate level. There will be those who are not enthralled with the idea of sharing power with an organization whose very purpose is to execute corporate strategy through projects instead of through functional departments. The director of the SPMO must be adept at building strategic relationships, and make it clear to others at his or her level that the SPMO exists to help the entire organization achieve its goals. By helping accomplish that, everyone will share in the rewards. Block suggests the following when dealing with the politics of projects:

- Assess the environment. Identify the key players and sources of power in the organization, and get to know them as individuals.
- Identify the goals of the key players. What are their overt goals, their covert goals. How can you help them achieve their goals?
- Assess your own capabilities. How are you at developing and nurturing personal relationships? What are your own personal values and how do they match up with those of other key players? Is there a conflict between the two? How good a communicator are you? (Communication skill is a prerequisite for the position of director of the SPMO).

Once this assessment is completed, Block says to continue with resolving key issues using the following familiar process:

- Identify the underlying problem to be solved or issue to be resolved.
- Develop alternative solutions for discussion with key players.
- Test and iterate solutions until final resolution is achieved.[9]

The Impact of Change on People

As functional organizations transition to project organizations, there will be a significant impact on the people of the organization. Change includes three distinct time periods: the *Ending*, the *Transition Period,* and the *New Reality*. Employees over these past years have become so immured to the multitude of changes taking place in all aspects of their lives that they are reacting in several ways, based on their own personal styles:

- Ignore it and it will go away.
- Embrace it immediately; whatever it is, it's new.

- Wait and see.
- Don't bother me with it, I'm busy.
- I'm too tired and discouraged to deal with anything new; I'm out of here.

In the majority of cases, it is seen as someone else's problem, and the employee feels powerless and disenfranchised.

A key factor to effective change is to provide a foundation of stability that forms the basis upon which all decisions are made, and then to apply specific strategies and procedures consistently when a change is implemented. Stability is provided by the vision and values of the organization; however, consistent implementation is provided through project management training throughout the organization.

There must be a clearly defined *Ending*, which communicates the reason the organization is moving to a project organization, the relationship between organizational goals and the proposed change, and specific steps for implementation, including each employee's involvement. The *Ending* cannot be ambiguous.

There must be understanding that a *Transition Period* will take place, expectations explained, and a time line presented and visible, with current updates when the *New Reality* will be in place. The fact that the company will not return to a *functional* structure must be expressed clearly and frequently. Awareness that employees will react in different ways makes it critical in this period for leaders to be more visible, more communicative, and more reassuring than at any other time. Opportunities for informal discussions, and transition and communication style assessment and training by outside consultants, can be extremely useful. Procedures for feedback, questions, and suggestions should be implemented and continued well after the *New Reality* is in place.

The 15-15-70 Rule

In any organization, there are a small percentage of people who are constitutionally incapable of dealing with change. Their strength is in providing continuity in a stable organization. Forcing them to undergo the disruption endemic in a dynamic organization is destructive to them and to everyone else in the organization. Therefore, it is critical to identify the change potential of employees and to make it possible for them to leave the organization in a positive manner.

- 15 percent of the employee staff will eagerly accept any change, just because it is new. They will lead the transition and thrive on the challenge because it is their nature to do so. In fact, their productivity and probability of staying with an organization long term will largely depend on the opportunity for new challenges. They are the Shifters and the Pioneers.
- 15 percent will be extremely uncomfortable with change. In a stable organization, they are often valued as Settlers, providing excellence by refining the details of the operations and ensuring consistency. They will do everything in their power to prevent change from upsetting these established routines.
- 70 percent range between these extremes and can be integrated successfully into the new system with proper coaching. The challenge of project managers and internal managers is to work closely to identify them and assist them in the transition.

Using Language to Create Community

Nearly 40 percent of the workforce is now contingent labor. It is projected to reach 60 percent early in this millennium. Therefore, less than half of the employee body will be "residents," the rest will be visitors, outsiders who move in and out, without ties to the company's long-term goals or to each other.

Language is a critical factor in establishing a sense of community, expressing its values, beliefs, and assumptions, and bonding people. It includes the industry/company terminology that is used to explain procedures and give instructions, the assumptions that are not verbalized, but nevertheless there (including those critical to safety), the implications that show the relative importance of what is being said, and simple casual conversation that makes people comfortable with each other.

If 60 percent of your workforce speaks numerous "foreign" languages, how can they be expected to function effectively? The answer is to integrate project management into the entire organization, using it to provide a common "language" and procedures. This should include:

- Companywide education in project management methodology.
- Introductory seminars for new employees and contingency workers to smooth the transition and immediately create a community that speaks the same language, rather than taking the risks inherent in multiple language translations.
- Encouraging managers to identify work that can be handled as projects and providing opportunities for each of their employees to gain experience in the methodology.
- Consciously incorporating project management language in all in-house communications.
- Providing a project partner for all outsourced projects, to work directly with the project manager on integrating the change into the organization.
- Including cultural change in the scope statements for projects managed internally and externally.
- Assuming that everyone in the organization is a stakeholder in every project and ensuring communication that is open, easily available, and constant.

You Made It: Signs of a Project Culture

How can you tell if you have achieved the goal of creating a project culture? One way is to reassess project management maturity using the same assessment tool originally used for the assessment of the need for change, as discussed above. An abbreviated method is to use the following checklist. An organization with all five of these elements up and running is well on its way to achieving success.

- A standard project management methodology, deployed throughout the organization and used by all project teams
- A meaningful, attractive career path for project managers
- Effective education, training, and certification for project managers, and training for team members, managers or project managers, and senior executives

- Ongoing support through a Strategic PMO at the corporate level
- A standard suite of software tools to support project managers

We might want to include a sixth element: integration of software systems throughout the enterprise, as discussed in Chapter 8. An integrated system provides project managers, managers, and other stakeholders with all the information necessary for real-time project planning, execution, and control. Software will not create a project culture, but it can support and reinforce the behaviors that make up a culture.

Notes

1. James S. Pennypacker, *Strategy and Projects: A Benchmark of Current Business Practices* (Glen Mills, PA: PM Solutions' Center for Business Practices, 2005).
2. Cheryl J. Walker and Jean Erickson Walker, "Transitioning Functional Organizations into Project Organizations," *Proceedings of the Project Management Institute Annual Seminars and Symposium*, Project Management Institute, Newtown Square, PA, 1999.
3. Benjamin Schneider, Arthur P. Brief, and Richard A. Guzzo, "Creating a Climate and Culture for Sustainable Organizational Change *Organizational Dynamics* (Spring 1996).
4. PMI Standards Committee. *A Guide to the Project Management Body of Knowledge,* 4th ed. (Newtown Square, PA: PMI, 2008).
5. Daniel Tobin, *Re-educating the Corporation* (Hoboken, NJ: Wiley & Sons, 1993): 235–242.
6. Nicholas Schacht, "Project Management Culture: An Anthropological Perspective," *PM Network* (September 1997): 53–56.
7. Walker and Walker, "Transitioning Functional Organizations into Project Organizations."
8. Patrick Brown, Sheila Grove, Richard Kelly, and Satyendra Rana, "Is Cultural Change Important in Your Project," *PM Network* (January 1997), 48–51.
9. Robert Block, *Politics of Projects* (Upper Saddle River, NJ: Prentice-Hall, 1983).

10
Knowledge Management and the PMO
Tracking Benefits and Learning from Experience

The PMO is the place where project management and knowledge management intersect. In fact, despite the recent flurry of excitement, software releases, and books about it, managing knowledge has *always* been an integral element of good project management. The processes for project closure, especially the capturing of lessons learned, are in effect knowledge management processes. However, going over what went well and what did not at the end of a single project does not necessarily harness that experience for the future good of the company. Knowledge management allows the enterprise as a whole to learn from the successes and failures of individual projects. To do this requires some central clearinghouse for those learnings—and the PMO is perfect for this role.

Another knowledge management role for the Project Management Office is that of tracking benefits derived from the initiatives (both strategic and tactical) in which the organization invests. Tracking benefits and measuring value has been a difficult evolution for project management, but, over the past decade, as project portfolio management processes have become more mature and information technology (IT) integration has rendered the organization more transparent, measuring the value of PMO efforts has come within reach.

Let's look at how knowledge management serves, and is served by, the PMO, both in terms of project management culture and performance measurement.

Knowledge Management: The Short Course

It may be helpful to begin this chapter by stripping some of the buzzword aura away from the term *knowledge management* (KM). As with many terms enjoying popularity in business circles, the definition can depend on whom you are asking—a software vendor, for example, will tend to give you a software-based answer. And sometimes knowledge management may go under another name, such as data mining, business intelligence, best practices sharing, or lessons learned collection and dissemination. We could make the case that all management is, in fact, knowledge management: the ability to keep, find, sort, and analyze the information we need to get the right things done in the right way.

To begin with, KM is both more and less than software. KM authority Wally Bock[1] suggests the following definition: "Knowledge management is the way that organizations create, capture, and reuse knowledge to achieve organizational objectives." He goes on to define KM as a four-part process that is iterative (Figure 10.1).

In the first step, people process data and information into knowledge. ("Knowledge is information combined with experience, context, interpretation, and reflection. It is a high-value form of information that is ready to apply to decisions and actions."[2])

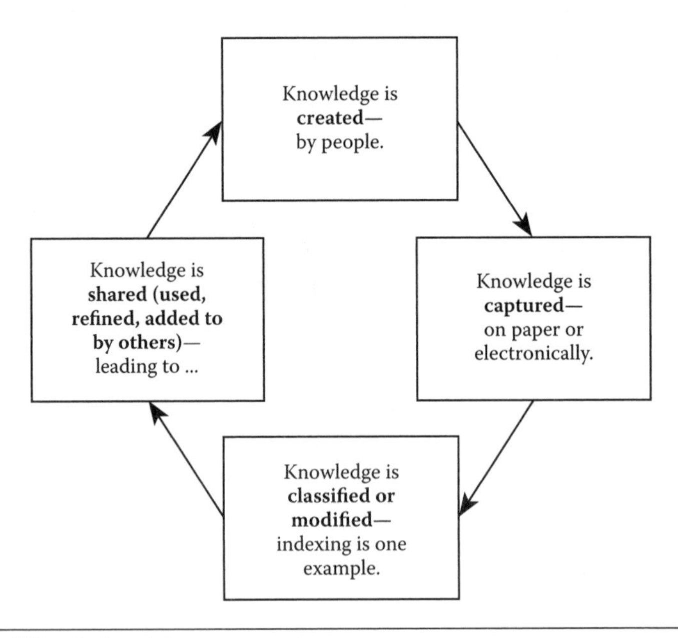

Figure 10.1 The knowledge management process.

In the second step, that knowledge is captured on paper or in a computer, or it may simply be remembered by the creators for reuse. In the third step, the knowledge is classified and modified, by indexing for example, or by the addition of keywords. This modification often adds context that makes the knowledge easier to reuse later. You can tell how well this step is being carried out by how easily people in the organization are able to find and use the knowledge when they need it. Finally, as knowledge is shared with others, it is refined, added to, and otherwise modified. This step merges into a fresh round of the cycle, as the new developments are captured, modified, and so on.

Companies initiate KM projects to find ways to capture, interpret, organize, disseminate, and capitalize on what they've learned; when these learnings take place on projects, project management and KM can merge and reinforce one another. Knowledge management and project management have been called *two revolutionary disciplines* that not only coexist but produce added value when they are married to each other.[3] And, of course, all the steps above have grown vastly easier as our IT infrastructure has evolved over the past decade.

Knowledge Repositories

Where and how is knowledge stored in your organization? There are three basic kinds of knowledge repositories, according to Bock. *Structured repositories* are databases, expert systems, and other storage that are searchable via indices, keywords, controlled vocabulary, and so on. *Unstructured repositories* include things like project reports, sales call notes, and other sources. These are searchable by free text means. These two kinds of repositories are for *explicit knowledge* (knowledge that's been captured and stored for future usage). However, *tacit knowledge*, which resides in the minds of people, is the most valuable. We all know and use the tools that are available to access this "repository"—phone directories, company directories, resource libraries, and other lists that show us who knows what. This is where social media like wikis really produce value, allowing many diverse and often distributed team members to collaborate on a solution or resource document, each one adding value from his or her own knowledge, experience, and perspective.

Key Success Factors in KM

Not surprisingly, a big part of knowledge transfer and management is communications planning, a skill that should be familiar to project managers. Below are some pointers offered by Bock and others:[4]

1. Limit your efforts to those things most likely to make a difference. This is simply another way of saying, focus on core competencies. What areas of project management might be most improved by better knowledge transfer in your organization? Some obvious areas might be in bringing new team members up to speed, or in stockpiling budget and actual figures so as to produce more realistic estimates.
2. Communicate, communicate, communicate. Give project personnel the widest possible array of information and knowledge: access to customer records, presentations, technical manuals, and other company, industry, and professional background and documentation, anything and everything they might need. The Web is the ideal tool for connecting people to information. You can give experienced people access to helpful tools and bring new personnel up to speed at a relatively low cost online. And, you can enlist them in improving what you put online to prevent it from becoming static. Encourage them to collect customer and technical lore and add it to the database (wiki, blog, social networking group site) that is available to help customer service and tech support solve problems quickly.
3. Give project personnel access to each other (see section The PMO as a "Community of Practice" below for more details on this). Again, the Internet is an ideal tool: listservs, chats, blogs, wikis, social networking sites, and discussion boards allow people dispersed across the company or the globe to share tips and experience. Personal contact is also important; we will discuss the many opportunities for project review meetings later in this chapter.
4. Identify the knowledge drains and gaps. Bock suggests surfacing key knowledge by asking: "What do we lose when key people leave?" or "What do we have to teach every new person?"

5. Facilitate exchanges among technical experts who have vast company knowledge stored in their heads; encourage them to become leaders in the project management "community of practice" within the organization to spread that knowledge around.
6. Use the resource library capabilities of project management software to catalog the competencies of employees to help match them up with the project teams that can best use their expertise, then expand these libraries beyond project management to encompass skills from other segments of the enterprise. Remember, all the great breakthroughs in science have come from "boundary-spanners"—people who bring insights from another, perhaps unrelated discipline. The same can be true in business.
7. Encourage best-practice sharing between projects, between departments, and between project managers within and outside your organization. Bock calls best-practice sharing "the most powerful knowledge management practice." It is also one of the simplest knowledge management tools; people who share an interest or a goal do it naturally. Companies can capitalize on this by providing infrastructure to spread the sharing around—networks, collaborative software, databases, libraries, and so on. However, the most important success factor in best-practice sharing doesn't involve technology. It involves strong leadership, with attention to culture and change. The application of technology can make a good process better, but will not be a substitute for organizational leadership that models an interest in knowledge management and process improvement.[5]

Table 10.1 offers a quick KM practices assessment tool for your organization.

Barriers to KM Success

The biggest barrier most organizations face in implementing effective knowledge management is a culture that rewards information hoarding. As noted above, software won't solve this problem; only a management culture that rewards knowledge sharing will solve it.

Table 10.1 The Knowledge Management Assessment

DIMENSION	QUESTION	SCORE
Leadership	Does your organization have a knowledge management strategy that is actively supported by the executive level, and which clearly articulates how KM contributes to achieving organizational objectives?	
Measures	Does your organization measure and manage its intellectual capital in a systematic way?	
Processes	Does your organization have systematic processes for gathering, organizing, exploiting, and protecting key knowledge assets including those from external sources?	
Explicit Knowledge	Is there a rigorously maintained knowledge inventory with a structured thesaurus or knowledge tree, and clear ownership of knowledge entities, that is readily accessible across the organization?	
Tacit Knowledge	Do you know who your best experts are for different domains of key knowledge, and do you have in place mechanisms to codify their tacit knowledge into an explicit format?	
Culture/Structure	Is knowledge sharing across departmental boundaries actively encouraged and rewarded? Do workplace settings and format of meetings encourage informal knowledge exchange?	
Knowledge Centers	Are there librarians or information management staff that coordinate knowledge repositories and act as focal points for provision of information to support key decision making?	
Exploitation	Are your knowledge and knowledge management capabilities packaged into products and services and promoted in your organization's external marketing?	
People/Skills	Have specific knowledge roles been identified and assigned, and are all senior managers and professionals trained in knowledge management techniques?	
Technical Infrastructure	Can all important information be quickly found by new users on your intranet (or similar network) within three mouse clicks?	

Note: Scoring key: Rate your organization (either the Project Office or the organization as a whole) on a score 0 to 10: 0 = doing nothing at all; 10 = world class KM. Scores will be more accurate if several people complete the assessment and average their scores.

Source: Adapted from Skyrme, D. (1999) *Knowledge networking: Building the collaborative enterprise,* Oxford, U.K.: Butterworth-Heinemann.

In a recent survey, Ernst & Young asked 431 U.S. and European firms about their knowledge management practices.[6] While 87 percent of respondents said that knowledge was critical to their competitiveness, 44 percent reported that they were poor or very poor at transferring knowledge within their organization. Respondents reported that the chief barriers included top management failure to express the importance of knowledge management (32 percent), lack of shared understanding of strategy or business model (30 percent), and organization structure (30 percent). These research results were echoed by a study co-produced by International Data Corporation (IDC) and *Knowledge Management* magazine. Their findings indicated that an organization's main KM implementation challenge stems from the absence of a "sharing" culture in the organization and employees' lack of understanding of what knowledge management is and what benefits it offers. To address these challenges, companies must make training, change management, and process redesign primary components of their KM initiatives.[7]

Before going on to discuss how to address some of these barriers, let's examine what existing project management practice has to offer in the way of KM.

Capturing Lessons Learned—and Beyond

In the spirit of "limiting your efforts to those things likely to make a big difference," the best place to begin focusing on the KM potential of a Strategic PMO is in that area of project management where process and technical information is gathered, the lessons-learned processes inherent to project management. The SPMO oversees the end of phase/end of process reviews of individual projects and guides project managers and the teams through the review process. In its role as manager of project management knowledge, the SPMO is responsible for documenting the results and conclusions from each end-of-phase review, for developing an action plan, and for preparing follow-up analysis to find if the things that were identified in the end-of-phase review are actually being implemented. On a macro level, the SPMO also should be in charge of capturing the knowledge created by every completed project and program and organizing it in such a way that this knowledge is readily available for future use.

A critical element of process management is continual improvement. This step can be incorporated within each of the project management processes (initiating, planning, executing, controlling, and closing) to ensure feedback on lessons learned and input of recommendations to improve the overall understanding, value, effectiveness, and efficiency of the organization's project management methodology.

Here's a quick overview of the processes for capturing knowledge that should be a part of any project management methodology put in place by the PMO.

First, there are two kinds of information about which we want to capture lessons learned:

1. Product of the project
2. Processes of the project

Product evaluation objectives should include comparing the final project against the original quality objectives, identifying problem areas, and determining how to address them. While it may seem logical to carry out a product evaluation when the product is complete (see section Final Product Evaluation below), many problems with quality can be avoided by including product evaluation elements in each end-of-phase review (see Chapter 5 for a discussion of quality assessment).

Process evaluation follows a similar evaluation procedure. A process end assessment (as part of a phase-end review) should be held as each of the phases in the project draws to a close, when the project manager and project team report their progress to the project steering committee, and to gain the approval to proceed with the next phase of the project. An assessment should normally be held at the end of every phase and every process in the project. This assessment reviews the overall progress of the project and the plan for continuing the project (Figure 10.2).

The objective of the process end assessment is to document the status and results of the current phase, prepare for the next process, and review the current phase results with management. Just as important, however, is to provide a summary of the work that was carried out in the phase and a record of the project and phase plans for use in

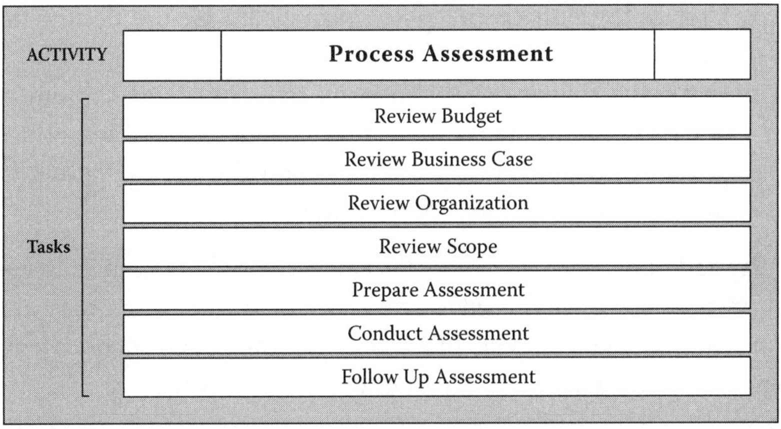

ACTIVITY | **Process Assessment**

Tasks:
- Review Budget
- Review Business Case
- Review Organization
- Review Scope
- Prepare Assessment
- Conduct Assessment
- Follow Up Assessment

Figure 10.2 Process assessment.

managing and controlling not only the current project, but any future project similar in nature. It also serves as a forum for issues identification, and a nexus of communication with the project steering committee.

Suggested steps for each phase end assessment include:

- Review the overall project budget and make changes based on the latest adjustments.
- Review project business case.
- Update the business case to reflect any changes in costs, benefits, and risks for the project.
- Review the resource requirements from the phase schedule and update the project organization accordingly.
- Review the latest statement of project scope and ensure that it still accurately reflects the current status and plans for the project.
- Refine the project plan for the next phase/process by adjusting the phase schedule and verifying that the criteria for success and completion of that phase/process are still valid.
- Decide who should attend the process end assessment in addition to the project steering committee and arrange a meeting to present the results of the assessment. At the meeting, request permission to proceed on the next process/phase of the project, or to stop the project if appropriate.

- Update the project and phase plan based on the decisions made by the project steering committee.
- Distribute minutes of the meeting to attendees and a summary to the project team. Document lessons learned from the phase and the planning process and provide process improvement recommendations to the PMO.

An additional step is that the project management process as a whole—the methodology—may need to be updated in order to capture learning from the process evaluation. This type of process assessment overview is best carried out as an integral part of the postproject review.

Project Closeout: The Knowledge Goldmine

All projects end at some point, that's the nature of project work. And though the project closing processes might seem like an afterthought, it is here that the organization can gain the maximum benefit from the work that was done.

All members of the project team and sponsoring committee will have valuable experience and feedback to be captured for future use. A large quantity of information is generated during a project, and this will have been stored with varying degrees of formality by the members of the team. This information needs to be filed away for possible future use. At one time, project information of this type would have gathered dust in binders on a shelf or in file cabinets. Now, however, thanks to knowledge management technology, these documents can remain live and readily accessible to the entire organization.

Project closure information is actually captured throughout the life of the project. It is important to capture information while it is still fresh in the minds of the participant. Consider what may be lost to the organization when people are transferred off the project before it ends, without having captured their key learnings from their participation. In the same way, subcontractors whose work only extends for a portion of the entire project timeline have input that can be valuable if processes are in place to record it.

The project should have been using lists and tracking mechanisms, such as change request logs, issues logs, a project blog, and other tools and templates that are part of your project management methodology (discussed in Chapter 5). These need to be closed to ensure all necessary work has been completed.

For smaller projects, the final project evaluation and postproject review steps can be very informal. The objective is to capture key learning and customer feedback.

Closing the project must take place in such a way that recommendations are identified that can be applied to future projects, mechanisms can be established for the continued development or improvement of the final project or product, the standard process and metrics (methodology) for this type of project can be improved, and that project management resources can be redeployed in the most beneficial way for their continued growth and for the company's best interests.

Postproject Review

Process improvement is the final step—and overall goal—of any quality process. This step evaluates the overall project management process itself and identifies any lessons learned from the project. If these lessons learned are likely to apply to future projects, they are provided to the PMO to be incorporated as revisions to the project management process. Be sure to focus on what worked well as well as what could have been improved

The objective of process improvement is to review the process and provide input to improve it. The process needs to be accomplished in a manner that involves both technical and business staff; covers the process used, the tools/techniques, deliverables, standards, and the organization; identifies things that were not necessary and additional phases, activities, and tasks that need to be added; and improves the quality and efficiency (time and cost) of future projects. Only then can the organization learn from this project and ensure improved performance in future projects.

Postproject review may take the form of a meeting, a facilitated workshop, a questionnaire, or a combination of the above. The review should involve all parts of the project organization and also may involve other staff personnel and key customer and supplier personnel. The postproject review collects all the data required for the Project Closure Report and any additional data required by the project.

To identify process improvements, review notes captured during the various process steps. Survey all team members, solicit their input, and thoroughly document each recommendation for process improvement, including the rationale for the suggested change/addition and the benefit to the company. Improvements should be focused on improving process efficiency and effectiveness. Classify the urgency of each change in order that the process steward can effectively prioritize improvements. Provide improvement recommendations to the process steward, and maintain a record of the recommendations as an audit trail.

Prior to the review session, conduct the Lessons Learned exercise using the template on the supplemental CD-ROM. Provide for detailed note-taking during the proceedings. Use the most up-to-date meeting technology available to allow everyone—the shy, the novice, and team member in India—to input thoughts and data. For the various topics, record the names of the commentators and the IT or business group they represent. After the session, complete the post-project review form by summarizing the responses of the attendees. Consider having a member of PMO or another third party facilitate the session.

Remember, the purpose of the review is to communicate and document the experiences of a project in order to better plan and manage future projects. Before finalizing the review summary, offer those who contributed a review draft of the proceedings so that they can correct their comments. (If appropriate, names may be withheld from the final summary.)

Why Are We So Bad at This?

All this sounds very well in theory, but research carried out by project management researchers Lynn Crawford at the University of Technology at Sydney (UTS) (Australia) and Terry Cooke-Davies of Human Systems, Ltd., in 1994[8] found that, while most project management groups are good at capturing lessons learned and storing them in some way, the step where KM fails for most project-oriented companies is in the *transfer* of that knowledge to future projects and project teams (Figure 10.3).

In Crawford's words, in terms of transferring lessons from one project to another, "a fault line runs down the center of the process."

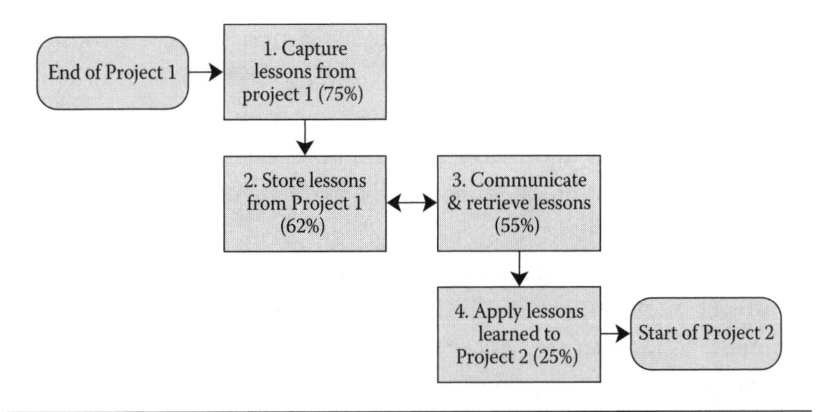

Figure 10.3 The knowledge gap in project management. Apapted from Crawford, L. et al. 1999.

Lessons learned are captured toward the end of a particular project and are the responsibility of an "outgoing" project team, whereas knowledge generation and transfer occur toward the beginning of a project and are the responsibility of an "incoming" project team. Some of the problem lies in the way project managers have traditionally thought of projects—as unique endeavors. "Unique" is even a part of the standard definition of project work. That being the case, it can be tempting to dismiss the experiences of Project A as being due to specific circumstances that don't apply in the circumstances of project B.

Another condition of project work is a sense of rivalry between teams; although Project B team has a high degree of respect for those who worked on Project A, they may be less than anxious to learn what that team has discovered. Crawford notes that such respect and sharing is much more likely if both project teams see themselves as members of a higher-level community, with accountability to all members of the community, rather than as separate entities in competition.

In addition, project managers tend to place the highest importance on aspects of the project other than knowledge transfer. The UTS researchers asked participants to rank the importance of project management practices relating to initiating and planning, monitoring and controlling, and closing and capturing lessons learned, and found that study participants ranked monitoring and controlling most highly, closely followed by planning, with closing and capturing lessons learned coming in a distant third.

Finally, in terms of personality characteristics, project personnel in the UTS study tended to be more task-oriented than people-oriented; in other words, they did not readily take time to share their knowledge with others. This is another area where new collaborative technology is helpful. "Techies" may more readily share thoughts on a Web site than in face-to-face meetings.

The PMO, as the central repository of project information, keeper of project management methodology, administrator of project management infrastructure, and trainer of project personnel, is in a unique position to maximize an organization's ability to learn from its own experiences. Software, discussed in Chapter 8, provides a partial answer to the resolution of the loss of knowledge over time, but "humanware," Lynn Crawford suggests, can be even more effective. Crawford's work with international networks of project management practitioners has led her to adopt the community of practice as a model way to nurture project management competence.

The PMO as a Community of Practice

> Management in a high-tech manufacturing organization performed an audit and discovered that design teams were repeating common mistakes, resulting in tens of millions of dollars in rework and repair costs. A simple solution was formulated: create communities of practice (CoPs) across design groups to enable project personnel to more easily capture and share valuable knowledge, information, and best practices. Using existing collaborative software tools and databases, the CoPs focus on linking product engineers together so that they can collaborate and get assistance in real time from experts across the organization. The CoPs filled the gap that sophisticated information systems could not; they allowed the engineers to ask each other: "What do I know that you need to know?" and "What do you know that I need to know?" These simple dialogues have resulted in significant savings and increased team satisfaction and innovation.
>
> *Embedding KM: Creating a Value Proposition*
> www.apqc.org, *May 2001*

"Any methodology worth its salt," says process improvement expert Michael Wood, "seeks to harvest the wealth of knowledge that exists within the minds of those who actually do the work."[9] How is this "harvest" best carried out? Wood contends that knowledge workers who are organized into cross-functional groups representing the end-to-end processes that actually deliver value to stakeholders can readily define and resolve the value gaps in the process.

The cross-functional group he describes is an apt description of the teams that accomplish project and program work in most organizations. Thus, an organization in which most work is accomplished by teams is already in a position to engage in continuous improvement of its own methodology and processes. All the PMO needs to do is learn how to nurture these cross-functional groups and provide them with the infrastructure to put up their "harvest" for safekeeping and future use. What organizational structures are best suited for making the most of that harvest?

Lynn Crawford has suggested that project management communities of practice can be invaluable in facilitating knowledge creation, transfer, and learning between individuals and across organizational boundaries.[10] Table 10.2 shows the qualities and functionalities that characterize a community of practice.

Table 10.2 Communities of Practice Characteristics

Common language	Group has a professional language of its own (jargon, terminology)
Shared background	Members have shared background or knowledge
Common purpose	The group has a common purpose that gives it an internal impetus
Creation of new knowledge	The work of the group and the interaction of the members create new knowledge for those members
Dynamism	Social distribution of knowledge takes place in the group
Evolution	Group develops beyond mere social interaction
Unofficial	Group evolves rather than being created
Voluntary	Membership is generally voluntary
Narration	Swapping war stories is a key way in which members share domain knowledge
Informal	Group is often informal—there is no hierarchy
Fluidity	Newcomers arrive and old-timers leave
Similar Jobs	Group members have similar jobs
Self perpetuating	As groups generate knowledge, they reinforce and renew themselves
Self-managing	Groups benefit from cultivation, but not from control

Source: Adapted from Hildreth, P. (2000) Communities of practice. Available at http://www-users.cs.york.ac.uk/~pmh/work.html#Communities (accessed April 7, 2000).

Communities of practice are not new, but the term has acquired new significance in the era of the knowledge worker and intellectual capital.[11] Communities of practice are groups of people informally bound together by shared interest and shared expertise—a simple enough definition and one that might apply to most voluntary groups in our society.

Yet, the importance of the community of practice to the development of technical and business knowledge began to be truly appreciated thanks to the Institute of Research on Learning (IRL), in Palo Alto, founded in 1987 and associated with Xerox's Palo Alto Research Center, and to the work of social learning theorist Etienne Wenger.[12] A key finding of the IRL's work was that learning is social. In this context, communities of practice have become associated with the concepts of the learning organization and knowledge management.

The primary reason to implement a PMO is to improve project performance across the organization. Standardizing on best practices is one way to accomplish that goal. Thus, the encouragement of best-practice sharing among all segments of the organization thus is a natural role for the PMO to take. By recognizing the shared interests and goals of project management practitioners throughout the organization, the PMO can play an important part in supporting, facilitating, and networking the informal communities that are "people repositories" of project knowledge around the enterprise.

Organizations, such as Boeing, NASA, NCR, and Ericsson, have provided support for project management communities of practice along these lines, even rewarding active participation in corporate communities of practice. Further, the development of communities of project management practice between organizations has led to initiatives that benchmark project management practices.[13]

Since it is a truism that informal channels of communication are faster than formal ones, and since "… A firm's competitive advantage depends more than anything on its knowledge … *on what it knows,* how it *uses* what it knows, and *how fast* it can know something new,"[14] the PMO's facilitation of project management communities of practice can contribute not only to the capture and growth of project management knowledge, but to the speed of innovation and change in the organization as a whole.

In addition, the community of practice, like a distributed computing network, links and expands the "repository" of project management knowledge that is stored in practitioner's minds—the tacit knowledge that is otherwise so difficult to manage.

What do the experts tell us about building and maintaining communities of practice? In "Knowing in Community," knowledge management expert Richard McDermott[15] listed a number of critical success factors for communities of practice:

1. Focus on knowledge important to both the business and the people. This echoes Wally Bock's advice to limit the focus of your KM efforts to areas where you can make a difference. The topics addressed by the community need to be ones about which people feel personally passionate.
2. Find a well-respected community member to act as coordinator, to keep people and create opportunities for people to share ideas. A coordinator need not be a leading expert, but must be someone who connects well with people.
3. Make sure people have time and encouragement to participate. Some companies specifically allocate a certain amount of project time to community activities to ensure that the time and energy people invested in the community would count in their performance appraisal.
4. Involve thought leaders who either have an important specialized knowledge or who are well-connected and influential members of the project management community. These people legitimate the community and draw in other members.
5. Make opportunities for contact. While documented reports, templates, tips, analyses, proposals, and so on are helpful to most community members, face-to-face contact is key to building a sense of commonality, enthusiasm, and trust.
6. Allow people to participate at their own comfort level. Even lurkers often get great value from a community where they can drop in quietly to find out who is working on what or learn about the field and make contact later.
7. Encourage real dialogue about real problems and cutting-edge issues by building a trustful and mutually helpful environment.

One of the things that these communities can do to add value in an organization is not only share, but benchmark best practices, both internally and against other organizations.

Benchmarking

In Chapter 5, we discussed the important role that benchmarking best practices plays in continuous improvement of process and methodology. It's in the PMO's interest to see that benchmarking takes place, both on a project level against other projects of that kind as well as against other organizations that manage similar projects. How else can the PMO offer proof that its policies are improving project results internally? How else can the organization determine if its project management practices are helping it to pull ahead of competitors?

Benchmarking provides an objective starting point for organizational improvements. It answers the question: "Where do we stand versus others?" and points toward new practices and new metrics; convinces middle managers of the need for change; identifies opportunities for improvement; and justifies the investments needed to realize these opportunities. As a tool for process improvement, it works best if it is iterative—part of an ongoing program.[16] The cyclical nature of projects provides a perfect organizational context for benchmarking efforts to remain dynamic and effective. Benchmarking process phases are described in Table 10.3.

High-performing organizations employ benchmarks and best practices as management tools to support strategic thinking and planning. They integrate benchmarking into the management of everyday processes and use networking and technology to expedite and optimize benchmarking. As project management knowledge collection and dissemination becomes more and more of an organizational priority, all levels of project personnel can learn to become receptive learners and skillful benchmarkers who actively practice innovative adaptation, and who recognize benchmarking as an essential skill.[17]

Benchmarking allows organizations a tool for gathering information and understanding the best practices external to their organization as well as internally. This information can allow organizations to improve profits/effectiveness, accelerate and manage change, achieve breakthroughs/innovations, and understand world-class performance,

Table 10.3 Benchmarking Best Practices

	PLANNING	ANALYSIS	INTEGRATION	ACTION
• **Benchmarking Process Phases**	• Identify benchmark subject • Identify benchmark partner • Determine data collection method • Collect data	• Determine competitive gap • Project future performance	• Communicate results • Establish functional goals	• Develop action plans • Implement plans • Monitor results • Recalibrate benchmarks
• **Benchmarking Critical Success Factors**	• Active management commitment to benchmarking	• Clear, comprehensive understanding of internal processes as a basis for comparison to industry best practices	• Realization that competition is constantly changing; future needs must be anticipated.	• Willingness to share information with benchmark partners
	• Focus on benchmarking first on industry best practices and second on performance metrics	• A concentration of other recognized leaders against which to benchmark	• Openness to new ideas; creativity and innovativeness in their application to existing processes	• A continuous, institutionalized commitment to benchmarking best practices

Source: Adapted from Camp, R.C. (1995). *Business process benchmarking*, Milwaukee, WI: ASQ Quality Press.

according to the American Productivity & Quality Center, which offers a benchmarking methodology to members. Some of the principles of APQC's methodology include:

- Make sure that the project is truly supported.
- Ensure that the scope of a benchmarking effort is very targeted to only the few key issues facing your organization or group.
- Make full use of the Internet, but make sure to verify all of the information found there.
- Examine your internal processes before starting a benchmarking project. Go through a true process mapping exercise with your internal peers.[18]

Once you have decided what features of your project management process to benchmark, you must first learn how the process under study works within your own company. The benchmarking team then learns how the process can be improved by visiting other companies, reading literature, and talking to employees. Employees who have been performing the process for many years are often the best sources for ideas.

The next step is to develop recommendations based on other companies' best practices and present them to management staff. Management buy-in will be crucial to the implementation campaign. Staff also should know the recommendations so that they can effectively participate.

From 2002 to 2009, PM Solutions' Center for Business Practices hosted a series of benchmarking events, in which PMO leaders from numerous leading companies came together to identify the challenges facing them in improving project management practice in their companies and share practices that worked well for them. The data generated from these informal benchmarking roundtables provided the basis for further research into the most pressing questions of PMO leaders, about strategic alignment, process maturity, resource management, PMO practices, and the like. What's notable is that the process itself was very simple and can be replicated wherever PMO staff and leaders congregate, either in person or on the Web. The sharing of information and ideas in this type of "action research" is knowledge management in action, and has the welcome side effect of building community relationships among coworkers and colleagues.

Tracking Performance: Knowledge Is Power

Numbers can tell a powerful story. Is a project successful? Is a new process working? Is our investment in training paying off? To ensure that your numbers tell the *right* story, you need to be consistent in the planning, collection, and analysis of these numbers. Thus, a good performance measurement system depends on KM practices.

Measurement provides you with the information necessary to make intelligent decisions about what you do. It can tell you how well you are doing, if you are you meeting your goals, if your customers are satisfied, if your processes are effective, if improvements are necessary, and, if so, in what parts of the organization.

A structured measurement program allows you to *identify* areas for performance improvement, *benchmark* against industry/competitors, *set targets*, *identify trends* for forecasting and planning, evaluate the effectiveness of changes, *determine the impact of project management* and *tell a story* about your organization's performance.

For companies that depend on the execution of projects for business success, project portfolio tracking is all about measurement. When accurate data is available, strategic decision makers can decide where to invest more, or less, effort. The data used to value each project must be accurate and current.

The same is true for those companies that expend large budgets on training. Is it making a difference? How do we know? What do we need to do to find out?

There are numerous ways to determine the performance and success of your programs; however, measures necessary to one organization may not apply to another. That's why it's important to plan for measurement, and to clearly understand the needs and goals of the organization in determining what to measure and why. The organization should seek to balance and streamline its measurement activities to ensure the measures it chooses will result in strategic organization-level improvements.

Where to Begin

First, if you start a measurement program, you want it to be successful. You need to realize that it takes time and commitment. There needs to be a clear understanding of what you are trying to accomplish, have senior management sponsorship, make sure employees are onboard with the project, decide upon a measurement framework to use, have someone accountable for the accuracy of the measures, have a sense of urgency for the results, and make sure you have the resources to gather and analyze the data.

The reason for measuring should always be about goals and objectives. You can never stop asking, "Why are we doing this?" "Value" is in the eye of the beholder, so the first step is to identify what's most important to the organization. When strategy is clear, this should be relatively easy. Common goals and objectives to measure include:

- Reducing costs
- Improving timing
- Improving quality
- Measuring the effectiveness of training
- Improving productivity

Once you start measuring performance, you can begin to start measuring value.

For example, improving schedule performance for all your projects over a period of a year can be translated into improvement in average project cycle time that can be translated into improvement in time to market, resulting in the number of new products your organization produces, which can add significant value to your organization through increased market share. Value measures, therefore, provide information on the performance of the organization rather than the performance of a project. They must be collected over a longer period of time (no more than quarterly) and over your portfolio of projects. This example also demonstrates how good measures should align with organizational objectives.

When schedule performance has been linked to increased market share, the value of training project personnel in scheduling becomes calculable. In this same way, it is possible to work backward from any strategic goal, drilling down to those measurable tasks that have an impact on goal achievement, and then developing training plans that directly impact those tasks. Using a measurement program of this type, the training function will always have a "hard" answer ready to the questions: Does this training pay off? And, by how much?

So, once you decide *what* you want to measure and *why*, the next question is: "Will the data for this measure be easily accessible?" You'll want to make the electronic capture of data a part of normal business operations. For example, time stamps on electronic documents or actuals reported through project plans and financial systems are captured through standard systems. This type of data collection reduces the amount of effort required to gather data and ensures a greater level of compliance. It also allows for ongoing access to information, allowing for the periodic reporting of metrics from the systems without the need for mobilizing a search team to track down information.

There are, of course, political implications to the collection of metrics, which is why strong sponsorship is required. One example of the kind of issue that will come up might be that, early on in SPMO execution, one of the critical elements is to be able to collect data relating to resources, budget, and schedule. An SPMO must have access to statistical data to validate whether or not you are working in accordance with the plan. There must be a collection of time reporting, cost reporting, technical adherence to deliverables, and other data. Taking the example of time reporting, which may seem simple enough, the issues surrounding time reporting reach into the human resources function and into the financial function. Setting up appropriate time-reporting capabilities for a project-oriented organization will require implementation of improved methods of process controls, procedures, timesheets, etc. In this case, IT will must be involved so that project teams can track time electronically. Integration with accounting/finance must be accomplished to collect actual time and costs from the financial system. The results of this data compilation and reporting must be compiled, consolidated, and regenerated into the appropriate formats for executive reporting and integrated into the project controls system for project-level status gathering, updating, and reporting. Integration of multiple financial systems, coordination among the various organizations, compiling the resulting data, and generating appropriate tailored reporting only begins to reflect the complexity of the integration challenges surrounding the myriad changes that an SPMO deployment entails.

A Model for Performance Measurement

Now that you have committed to a measurement program, have appropriate sponsorship, and know what your ultimate objectives are, you will need a model to begin implementing. The PEMARI model established by PM Solutions has proved to work well in dozens of organizations. This model integrates a number of processes:

- Planning. A process for understanding key success factors, identifying stakeholders and roles and responsibilities, identifying performance management goals, and developing a program plan.
- Establishing metrics. A process for identifying and selecting performance measures, developing measurement scorecards (high-level measures defined at the governance level; specific metrics that roll up into these identified at the departmental or program level).
- Measurement. A process for planning for data collection, including data source and information technology required; collecting data and ensuring data quality (a joint responsibility of IT and the Strategic PMO as owner of the portfolio processes).
- Analysis. A process for converting data into performance information and knowledge, analyzing and validating results, and performing benchmarking and comparative analysis (a joint responsibility of IT and the SPMO).
- Reporting. A process for developing a communications plan and communicating performance results to stakeholders (a responsibility of internal communications).
- Improvement. A process for assessing performance management practices, learning from feedback and lessons learned, and implementing improvements to those practices (a joint responsibility of the SPMO as portfolio owner, and executives responsible for governance).[19]

Developing Performance Measures

While there is general agreement that "you can't manage what you can't measure," the actual measurements themselves usually prove to be a source of conflict. What are we measuring, and why? What *should* we be measuring? What's the connection between the performance measures we collect regarding individuals and their tasks and the ultimate performance of the company—if any? And what, in reality, does "performance" mean, on an organizationwide scale? Is it merely making money? And, if so, how much? Measures are the easy part, knowing what you want to measure, and why, is hard.

Therefore, following a structured process helps to develop less fuzzy measures, while engaging a wide variety of the people who actually do the work in the process prevents the chosen measures from diverging too far from reality or ease-of-use.

There is no single set of measures that universally applies to all companies. The appropriate set of measures depends on the organization's strategy, technology, and the particular industry and environment in which they compete. Like any aspect of any "living company," measures cannot be static, they cannot be chosen once and locked into place. Along with strategy, they evolve and are refined as the organization becomes more focused on and skilled at meeting strategic goals.

Measurement Planning

Planning your performance measurement program begins with identifying the measurement program team and their roles and responsibilities, and defining measurement program goals. Next, identify what, if any current performance measurement systems are in place. What will be your implementation approach? Like any program, a measurement program needs a program plan and a clear understanding of terminology among the team. Some suggested roles on the measurement team include:

- Sponsor
- Representatives from stakeholders
- Project manager
- Data collection coordinator
- Data analyst
- Communications coordinator
- Measurement analyst

Establishing and Updating Measures To develop a list of potential measures, start by brainstorming all the possible measures that would be meaningful to the goals you are trying to achieve. Measures flow from goals and objectives and are developed collaboratively with the measurement team and stakeholders. Explore the inventory of common measures (see Table 10.4 for examples) and follow these criteria for effective measures:

- Does this measure provide meaningful information?
- Is it supported by valid, available data?
- Is it cost effective to capture?

Table 10.4 Top Ten Measures

Return on Investment. (Net Benefits/Costs) × 100. This is the most appropriate formula for evaluating project investment (and project management investment). This calculation determines the percentage return for every dollar you've invested. The key to this metric is in placing a dollar value on each unit of data that can be collected and used to measure Net Benefits. Sources of benefits can come from a variety of measures, including contribution to profit, savings of costs, increase in quantity of output converted to a dollar value, and quality improvements translated into any of the first three measures.

Productivity. Output Produced/Unit of Input. Productivity measures tell you whether you're getting your money's worth from your people and other inputs to the organization. A straightforward way to normalize productivity measurement across organizations is to use revenue per employee as the key metric. Dividing revenue per employee by the average fully burdened salary per employee yields a "productivity ratio" for the organization as a whole. Other productivity metrics might be the number of projects completed per employee or the number of lines of code produced per employee. The key to selecting the right productivity measures is to ask whether the output being measured (the top half of the productivity ratio) is of value to your organization's customers.

Cost of Quality. Cost of Quality/Actual Cost. Cost of quality is the amount of money a business loses because its product or service was not done right in the first place. It includes total labor, materials, and overhead costs attributed to imperfections in the processes that deliver products or services that don't meet specifications or expectations. These costs would include inspection, rework, duplicate work, scrapping rejects, replacements and refunds, complaints, loss of customers, and damage to reputation.

Cost Performance Index. Earned Value/Actual Cost. The CPI is a measure of cost efficiency. It's determined by dividing the value of the work actually performed (the earned value) by the actual costs that it took to accomplish it. The ability to accurately forecast cost performance allows organizations to confidently allocate capital, reducing financial risk, possibly reducing the cost of capital.

Schedule Performance. Earned Value/Planned Value. The schedule performance index is the ratio of total original authorized duration versus total final project duration. The ability to accurately forecast schedule helps meet time-to-market windows.

Customer Satisfaction. Scale of 1 to 100. Meeting customer expectations requires a combination of conformance to requirements (the project must produce what it said it would produce) and fitness for use (the product or service produced must satisfy real needs). The customer satisfaction index comprises hard measures of customer buying/use behavior and soft measures of customer opinions or feelings, weighted based on how important each value is in determining overall customer satisfaction and buying/use behavior. It includes measures such as repeat and lost customers, revenue from existing customers, market share, customer satisfaction survey results, complaints/returns, and project-specific surveys.

Table 10.4 Top Ten Measures (Continued)

Cycle Time. There are two types of cycle time: project cycle and process cycle. The project life cycle defines the beginning and the end of a project. Cycle time is the time it takes to complete the project life cycle. Cycle times for similar types of projects can be benchmarked to determine a standard project life cycle time. Measuring cycle times also can mean measuring the length of time to complete any of the processes that comprise the project life cycle. The shorter the cycle times, the faster the investment is returned to the organization. The shorter the combined cycle time of all projects, the more projects the organization can complete.

Requirements Performance. To measure this factor you need to develop measures of fit, which means the solution completely satisfies the requirement. A requirements performance index can measure the degree to which project results meet requirements. Types of requirements that might be measured include functional requirements (something the product must do or an action it must take) and nonfunctional requirements (a quality the product must have, such as usability, performance, etc.). Fit criteria are usually derived some time after the requirement description is first written. You derive the fit criterion by closely examining the requirement and determining what quantification best expresses the user's intention for the requirement.

Employee Satisfaction. An employee satisfaction index determines employee morale levels. The ESI comprises a mix of soft and hard measures, each assigned a weight based on their importance as a predictor of employee satisfaction levels; for example (percentage represents weight): climate survey results (rating pay, growth opportunities, job stress levels, overall climate, extent to which executives practice organizational values, benefits, workload, supervisor competence, openness of communication, physical environment/ergonomics, trust) (35 percent), focus groups (to gather in-depth information on the survey items) (10 percent), rate of complaints/grievances (10 percent), stress index (20 percent), voluntary turnover rate (15 percent), absenteeism rate (5 percent), and rate of transfer requests (5 percent).

Alignment to Strategic Business Goals. Most project management metrics benchmark the efficiency of project management—doing projects right. You also need a metric to determine whether or not you're working on the right projects. Measuring the alignment of projects to strategic business goals is such a metric. Survey of an appropriate mix of project management professionals, business unit managers, and executives. Use a Likert scale from 1 to 10 to rate the statement: Projects are aligned with the business's strategic objectives.

Source: Crawford, J.K., Cabanis-Brewin, J., and Pennypacker, J.S. (2007) *Seven steps to strategy execution.* Glen Mills, PA: PM Solutions' Center for Business Practices. With permission.

- Is it acceptable to stakeholders?
- Is it repeatable? Actionable?
- Does it align with organizational objectives?

Next, prioritize and select a critical few measures, keeping the number of measures at each management level to a minimum. A few criteria for prioritization of measures might include their importance to the execution of goals, the ease of accessing the data, and the ease of acting to change the performance.

This process leads to the development of a scorecard of vital measures, with each measure clearly defined as a "measure package" that details the *what, why, when, who*, and *how*:

- What is the measure?
- Why do we measure it?
- How will the data be captured?
- When will the data be captured?
- Where does this information reside?
- Who is the process owner for this data?

Address reliability, timeliness, and accuracy issues for each measure. One hurdle you may run into early is whether your information systems are designed to support the kind of data collection and reporting that will be most useful to your goals. Data collection should be:

- Focused on the organization's assessment and improvement needs
- Flexible to take advantage of any data source or method that is feasible and cost efficient
- Simple and aligned with the organization's needs to provide clear, relevant information
- Consistent to allow comparisons and easy transition from one data set to the next

Analyzing the Data Use a scorecard method to organize and aggregate the data, grouping measures by their relationship to key organizational areas of concern, such as financial measures, customer satisfaction measures, process measures, and employee satisfaction measures. In order to analyze and validate results, you must formulate precise questions that you are trying to answer (e.g., "Where is the most costly resource bottleneck in the organization?"). Then, collect and organize the data and facts relating to those questions. Analyze the data to determine the fact-based answer to the questions, and present the data in a way that clearly communicates answers to the questions. Any measurement program that fails to provide answers to the questions it raises, from an executive buy-in point of view, is doomed.

To validate your results, ask these questions:

- How does actual performance compare to a goal or standard?
- If there is significant variance, is corrective action necessary?

- Are new goals or measures needed?
- How have existing conditions changed?

Perform benchmarking and comparative analysis as needed to establish performance in comparison to competitors and companies or organizations that have a similar focus, or to determine whether the performance of the organization is improving.

Continuous Improvement Measurement, like any organizational improvement initiative, cannot be done just once. Having baselined your measurements, you will have to iteratively measure in order to develop trends. In addition, review for factors that impact the program. Some common barriers to success include inappropriate members on the measurement team, changes in organizational strategies and objectives, new measurement best practices, and feedback and lessons learned from your baseline efforts. Continually monitor these barriers to success, helping to improve understanding of the measurement process, of the organizational strategies and objectives, of your key business processes, and stakeholder needs. Regular communication will maintain senior management and employee involvement. Linking the measurement program to a system of accountability for results creates a sense of urgency and relevance.

Piloting the Measurement Program

Implementing a new measurement program or measurement improvement initiatives, such as new data collection methods, new scorecards, or new performance measurement information systems, can all benefit from the use of pilots.

- For a new measurement program, execute a pilot that involves ad hoc collection of data for a few select measures. This means no data collection processes are established. Also collected are characteristics of the measures, such as data quality and collection difficulty.
- Deliver a draft version of a scorecard to an appropriate audience.
- Collect feedback from data sources, audience, and measurement team members.

- Formally document what was learned from the pilot. Focus on issues that will provide information about the measurement process and how to improve it.
- Depending on the results, launch a new pilot, or launch a new or more robust measurement program.

Some Cautions about Doing Internal Research Projects

If you are seeking to prove the value of a PMO, either by collecting external research to make your case (benchmarking, for example), or by creating internal research to justify your investments, here are six important points to keep in mind.

1. What's in a name? The term *Project Management Office* suffers from having a welter of definitions. To one company, it's a single person on a mission to promote a methodology. To another, it's a bank of project mentors that fan out across the enterprise to teach people how to fish. In some large capital project contractors, a PMO might be a kind of war room for a single project. So, you first have to find out which PMO paradigm the participants are coming from, even before discussing the topic.

This all-over-the-map character isn't a bad thing; it's an essential part of the nature of a concept under development. While companies figure out how to make a concept into a set of workable practices, we can expect these practices to take many forms. However, in one way, the concept itself can be hurt when in-practice PMOs are so variable. In the project management field, there is limited research being done, so whatever studies are completed tend to be widely quoted (and sometimes misquoted; see no. 4 below). One bad piece of information (an erroneous assumption drawn from confusing results to a badly designed questionnaire) can have immense impact.

So, if you are creating a survey about PMOs, be sure to give the respondents a way to indicate what PMO means to them. A checklist of types of PMO organization with "check as many as apply" is one way to do this.

Another related question might be length of time since implementation. A one-person, brand-new PMO cannot possibly have had time to have much of an effect on organizational factors like career planning, portfolio success, and the like. Always offer the "not applicable" option in multiple choice questions.

2. The question makes the answer. In one recent PMO study I've seen, the results indicated that, for the majority of responding companies, those with a PMO had higher project failure rates. There are a couple of ways to explain this result. The companies surveyed might have been really bad at PMO implementation, but that seems unlikely. More likely is that the question didn't elicit a description of reality. For example, in many companies no one is keeping track of project failure rates. Therefore, the known failure rate is zero. Along comes the PMO and begins collecting metrics. Aha! Project failure rate is, say, 15 percent. Does this mean failures went up? No, but that's what it will look like if all you ask is pre-PMO and post-PMO failure rates.

Survey design is both art and science and a poorly designed questionnaire can obfuscate more than it illuminates. It would have been good to know things about these companies, such as:

- Did they take a baseline measurement of failures before they implemented the PMO?
- How long have they been tracking failures?
- What failure prevention measures have been undertaken by the best performing companies in the study? By the worst performing?

The problem, of course, is that the more precise your questions are the fewer of them you can ask. It's a law of questionnaire development that few people will respond to a survey questionnaire that comprises more questions than will fit on about one page. People have to be highly motivated to respond to longer surveys.

3. Data becomes valid—or meaningless—in the analysis phase. Especially if there are any qualitative questions (open-ended questions that elicit individualized answers), the task of analyzing survey results belongs not just to number crunching, but to creative, critical thinking. Good numbers don't always translate into sound conclusions. That's why you should "go to the source."

4. Go to the source. What is the source of the statistic in question trying to sell you? On the surface, it may seem like disinterested information, but in today's society, we are frequently the targets of marketing disguised as information. When collecting research studies, go to the

source instead of relying on quotes. I've recently seen some of our own findings misrepresented in press releases, and from a university, at that. More than that, look at the numbers yourself, not just at the packaged results offered in an executive summary. Your knowledge of your own industry may allow you to draw conclusions from the data that the report writer missed.

5. Remember that even the best surveys raise more questions than they answer. Otherwise, research would have ended a long time ago. The thoughtful researcher—or consumer of research—won't accept as final any results that seem counterintuitive. When I see an assertion such as "instituting career paths has no effect on project success rates," I immediately wonder: Why not? How long have they been in place? Do they offer technical and nontechnical tracks? Is professional development cost supported by the organization? Are project managers more interested in other aspects of reward rather than promotion or advancement? This is why articles in scholarly journals normally end with a section listing all the further research that needs to be done to explain the results of their research.

6. When doing internal research, be mindful of the Hawthorne effect. Merely by the way you structure your metrics gathering, you can alter the processes and outcomes that create that data.[20]

Conclusion

All companies compete based on knowledge. We leverage knowledge to improve processes, serve customers, update operations, and bring products to market. Many companies mistakenly believe that investments in technology are the primary route to success, while ignoring the deep well of knowledge and expertise that exists within the organization's members. Knowledge management techniques, when used to capture and link the experiences of project personnel, can offer a way to maximize resources. And further improving processes through benchmarking raises the bar another level. Research has show that benchmarking results in more efficient processes, which can generate substantial cost savings, as much as range from 15 to 45 percent, according to some experts.[21] In addition, from our *Strategy and Projects* research, we know that high-performing organizations are significantly better than average at having performance measures established at each organizational level (business unit, portfolio, program, project) to link up with the strategic performance expectations of the entire company.[22] By combining the principles of knowledge management, project management, and performance management, the organization can make the most of what it knows, learn more from its projects, and leverage knowledge gathered about competitors and industry.

Notes

1. Wally Bock, "Knowledge Management 101," *Intranet Journal*, http://idm.internet.com/articles (accessed May 2000). (The article originally appeared in *Bock's Briefing Memo* newsletter at www.bockinfo.com.)
2. T. Davenport, D. De Long, and M. Beers, "Successful Knowledge Management Projects," *Sloan Management Review* (Winter 1998).
3. N. Olonoff, "Knowledge Management and Project Management," *PM Network* 14, no. 2 (2000): 61–64.
4. Cinda Voegtli, "Know-All 10," archived on www.gantthead.com (accessed May 2000).
5. Bock, "Knowledge Management 101"; Voegtli, ibid.
6. Dave Webb, "Corporate Culture Blocks Better Use of Knowledge," *Computing Canada* (September 1, 1998).
7. State of Knowledge Management, *Knowledge Management* (May 2001).
8. Lynn Crawford and Terry Cooke-Davies, "Enhancing Corporate Performance through Sustainable Project Management Communities," paper presented at the *Proceedings of the 30th Annual Project Management Institute Seminars and Symposium*, Newtown Square, PA: PMI, 1999.
9. Michael Wood, "What Is a Process Improvement Methodology Anyway?" www.gantthead.com/articles.
10. Lynn Crawford and Terry Cooke-Davies, "Managing Projects—Managing Knowledge: Sharing a Journey towards Performance Improvement," paper presented at the *International Project Management Association Conference*, Cairns, Australia, 2000.
11. T.A. Stewart, *Intellectual Capital: The New Wealth of Organizations* (New York: Doubleday, 1999).
12. Etienne Wenger, *Communities of Practice* (Cambridge, MA: Harvard Business School Press, 1991); E.C. Wenger and W.M. Snyder, "Communities of Practice: The Organizational Frontier," *Harvard Business Review* 78, no.1 (2000): 139–146.

13. Crawford and Cooke-Davies (Enhancing Corporate Performance).; also Frank Toney and Ray Powers, *Best Practices of Project Management in Large Functional Organizations* (Newtown Square, PA: PMI, 1997).
14. L. Prusak, *Knowledge in Organizations* (Boston, MA: Butterworth-Heinemann, 1997): ix.
15. Richard Mcdermott, "Knowing in Community: Ten Critical Success Factors in Building Communities of Practice," *Knowledge Management Review* (May/June 2000).
16. Jeffery C. Egan, "Benchmarking as a Change Agent at IBM," *Supply Chain Management Review* (Winter 2000).
17. Christopher E. Bogan and Michael J. English, *Benchmarking for Best Practices* (New York, NY: McGraw-Hill, 1994).
18. "Embedding KM: Creating a Value Proposition," American Productivity and Quality Council, www.apqc.org (accessed May 2001).
19. Deborah Bigelow Crawford, "Mastering Performance Measurement" (white paper, PM College, 2008).
20. Jeannette Cabanis-Brewin, "Lies, Statistics, and the PMO: Some PMO Research Compares Apples with Oranges and Gets Fruit Salad," Developer.com: http://www.developer.com/mgmt/article.php/3399851/Lies-Statistics-and-the-PMO.htm (accessed August 26, 2004).
21. Kimberly Lopez, "How to Measure the Value of Knowledge Management," *Knowledge Management Review* (March/April 2001).
22. *Strategy and Projects: A Benchmark of Current Business Practices.* Center for Business Practices, 2005.

Appendix A: The State of the PMO 2007 to 2008

A Benchmark of Current Best Practices

PMOs: Growth and Expansion

Over the past decade, the Center for Business Practices has been involved in gathering data on project management trends from survey research, literature research, and from action research, such as our Project Management Benchmarking Forums. In that time, we have seen a steady vertical climb in the indicators of organizational influence for project managers and project management. Nowhere is this increased influence more notable than in tracking the prevalence and roles of the Project Management Office.

For example, in our 2000 Value of Project Management study, only 47 percent of the respondents had implemented a project office of any type. By 2006, 77 percent of the respondents to our Project Management: The State of the Industry survey had implemented PMOs; of those, 35 percent had an enterprise-level (or "strategic") PMO. This year, 54 percent of the respondents reported having an enterprise-level PMO in place. Even factoring in the differing research objectives of these studies, the upward trend is unmistakable, both in sheer numbers of PMOs and in the rising organizational clout.

Yet, questions remain. Top project management leaders in many companies are still struggling with verifying the value of their PMOs. The results from this study suggest that merely implementing a PMO is not a panacea. Instead, it is PMO *maturity* that makes a difference to the organization. Our data suggests, as PMOs become more mature, organizational success metrics improve. In addition, the mature PMO takes on more roles in both portfolio management and people management, thus elevating its value to the organization.

Key Findings

PMO Maturity PMO Maturity is rated on a scale from Level 1 to 5 (immature, established, grown up, mature, best in class).

- There is a strong correlation between organizational performance and the maturity of PMOs. Mature PMOs show significant improvement in organizational performance. "Performance improvement" is defined as rating higher on a scale of 1 to 5 on how well the organization performs in the eight measures of performance listed in the chart shown (only Levels 1 to 3 are listed because too few Level 4 to 5 PMOs responded to draw accurate conclusions).
  - There is a strong correlation between level of PMO maturity and organizational performance. Organizations with PMOs show significant improvement at each level of PMO maturity.
  - 6.2 percent overall performance improvement from PMO Level 1 to Level 2
  - 14.6 percent overall performance improvement from PMO Level 2 to Level 3
  - 10.5 percent overall performance improvement from PMO Level 3 to Level 4
- Just having a PMO does not lead to performance improvement. It isn't until the PMO becomes more mature that improvement occurs, but that improvement is steady and significant.

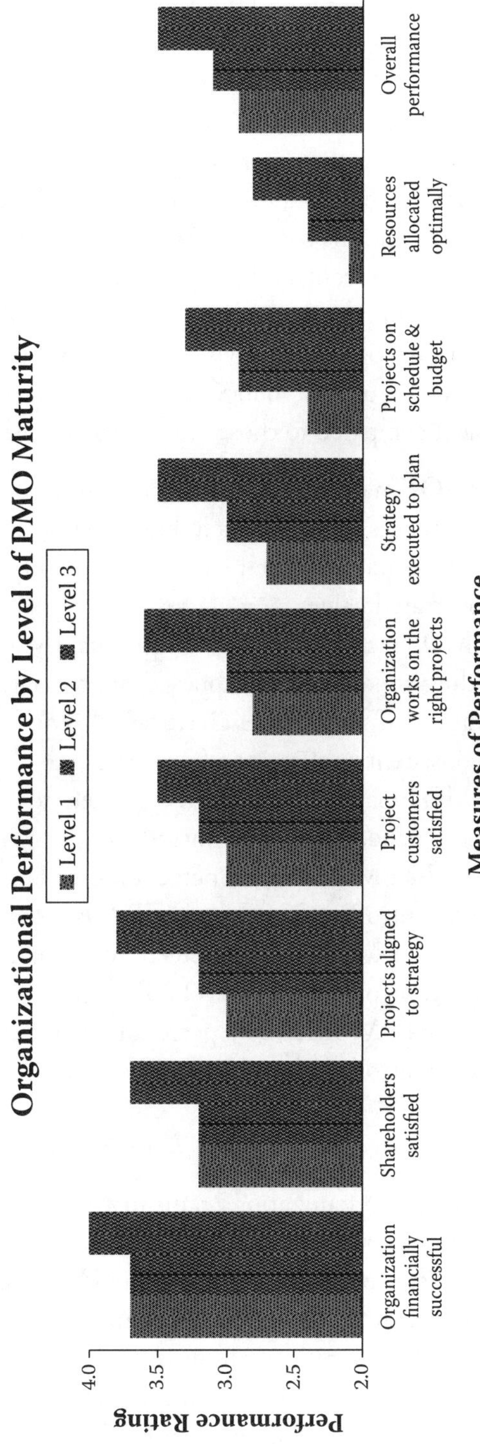
Organizational Performance by Level of PMO Maturity
Level 1
Level 2
Level 3
Performance Rating
4.0
3.5
3.0
2.5
2.0
Organization financially successful
Shareholders satisfied
Projects aligned to strategy
Project customers satisfied
Organization works on the right projects
Strategy executed to plan
Projects on schedule & budget
Resources allocated optimally
Overall performance
Measures of Performance

    - Organizations with PMOs at Level 2 maturity and higher show 3 percent overall performance improvement compared to those organizations with no PMO.
    - Organizations with PMOs at Level 3 maturity and higher show 11 percent overall performance improvement compared to those organizations with no PMO.
    - Organizations with PMOs at Level 2 maturity and higher show 8 percent budget/schedule performance improvement compared to those organizations with no PMO.
    - Organizations with PMOs at Level 3 maturity and higher show 16 percent budget/schedule performance improvement compared to those organizations with no PMO.

- As PMOs mature, they are significantly better at meeting critical success factors, including having effective sponsorship, accountability, competent staff, quality leadership, and demonstrated value.
- As PMOs mature, they have significantly fewer challenges, including stakeholder acceptance, appropriate funding, demonstration of value, role clarification, conflicting authority, and consistent application of processes.
- As PMOs mature, they are more likely to staff professional planners, schedulers, and controllers.
    - Level 2 PMOs have 14 percent more planners, schedulers, and controllers than Level 1 PMOs.
    - Level 3 PMOs have 24 percent more planners, schedulers, and controllers than Level 2 PMOs.
    - Level 4 PMOs have 70 percent more planners, schedulers, and controllers than Level 3 PMOs.

High-Performing Organizations versus Low-Performing Organizations

High-performing organizations rank in the top 25 percent in overall organizational performance based on ratings in eight measures of performance (strategy execution, shareholder satisfaction, customer satisfaction, budget/schedule performance, financial performance, resource allocation, strategic alignment, and portfolio performance). Low-performing organizations rank in the bottom 25 percent. (See page 43 of the Research Survey, “On Performance and Value.”)

- High-performing organizations are more likely to have an enterprise PMO (EMPO) (65.8 percent of high-performing organizations have EPMOs compared to only 48.6 percent of low-performing organizations).
- PMOs have been in place 29 percent longer in high-performing organizations (4.5 years) than in low-performing organizations (3.5 years).
- High-performing organizations are 30 percent more likely to have steering committees (64.9 percent) than low-performing organizations (52.1 percent)
- PMO functions performed significantly more by high-performing organizations include strategy formulation, portfolio risk management, benefits realization analysis, contract preparation, outsourcing, project opportunity process development, resource assignment process development, management of a staff of project planners/controllers and business relationship managers, and resource identification and optimization.
- Low-performing organizations are impacted significantly more by challenges than high-performing organizations, including conflicting authority (36 percent more), role splitting with competing authority (31 percent more), roles not clearly delineated (42 percent more), lack of a project management career path (24 percent more), inadequate opportunities for professional development (25 percent more), and inadequate project management skills (31 percent). The respondents were asked to rate the impact of the resource challenges listed in the chart above on their PMOs on a scale of 1 (not at all) to 5 (to a great extent).
- High-performing organizations outsource 135 percent more often than low performing organizations.
- High-performing organizations evaluate project manager and team competency significantly more often than low performing organizations.
- There is no correlation between project managers reporting to the PMO (as opposed to them just being supported by the PMO) and organizational performance.
- High-performing organizations have larger PMOs (30 percent more staff) and rely on more specialized roles (they have more staff performing those roles), including mentors (136 percent

increase), team leads (467 percent increase), planners (147 percent increase), controllers (116 percent increase), and relationship managers (698 percent increase).
- PMOs at high-performing organizations are 66 percent more mature than at low-performing organizations (average level of maturity is 2.9 vs. 1.7).

General Observations

- Organizations average (median) 31 projects per year. Organizations with a PMO work on more projects per year (38) than those without a PMO (18).
- PMO budgets range from $0 to $50 million a year (with a median average of $600,000). The PMO budget is on average, 1.7 percent of the organization's budget (median).
- The average PMO has 8.0 people reporting to it (the range of PMO size is huge, from a single person to more than 100). Also, older PMOs have significantly more staff (those 5+ years old average 16.5 staff vs. 8.0 overall).
- Fully 70 percent of PMOs have training goals for their staff. And PMOs in high-performing organizations are far more likely than those in low-performing organizations to have training goals (79.7 vs. 60.6 percent).
- The top two issues PMOs struggle with are (1) forecasting the need for resources and (2) resolving resource conflicts.
- Governance issues top the list of PMO challenges: Companies lack the compliance structure to make project management processes consistent throughout the organization, and project leaders still labor under conditions where responsibility and authority are not allied. However, as PMOs age and mature, they have fewer challenges and are significantly better at meeting all challenges listed.

Note: The Center for Business Practices is now PM Solutions Research. This survey was updated in May, 2010, and is available at this writing at www.pmsolutions.com/insights.

Appendix B: Selected PMO of the Year Award Winners (2007 and 2008)

2007: Norton Healthcare Enterprise PMO

Healthcare Projects: A Serious Business

Every project manager feels that, if the project doesn't go well, it will be "a disaster." But in the healthcare field, a malfunctioning piece of software, incomplete records, or poorly trained resources can literally mean the difference between life and death for the patients who depend on the myriad projects, both on the IT and clinical side, that are carried out daily in a large hospital chain. That's why Norton Healthcare's EPMO is serious about making sure that all the projects they manage are completed on time, within budget, and with a high degree of quality. More than that, the EPMO has taken the lead in introducing portfolio management methodology as a companion to strategic planning, to make sure the right projects for the health of the nonprofit are being selected.

With five hospitals, two medical centers, eight immediate care centers, 35 physicians' practices, and more than 2,000 doctors and 3,000 nurses serving 40 locations across the region, Norton is Kentucky's largest health care provider. The organization cares for over 1 million patients a year including 62 percent of all babies delivered in the region and 51 percent of all emergency care. With these kinds of patient care numbers, it is not surprising to learn that Norton Healthcare is also one of the area's largest employers, with a payroll of more than $340 million. Despite national heath care worker shortages, Norton's turnover is down by 42 percent, and the company has won numerous HR-related awards, including "Best Place to Work in Kentucky" (March 2006).

Best in Class Practices

Efficiency. The Norton Healthcare EPMO is a small group supporting a large organization, and one in which there are a variety of organizational "subcultures," from nursing to HR to marketing/communications to IT. In this complex environment, says Yackey, the key is "to operate as efficiently as possible or go down holding hands and singing 'Kum Ba Yah.'" Thus, the EPMO has striven for efficiency and user-friendliness in establishing project management throughout the enterprise. They have standard document templates for all of their services and use documentation from other similar projects to avoid reinventing the wheel.

Cultural Sensitivity. "We had to evaluate the culture as to the degree of change they could handle, while introducing better, more effective ways of conducting business. We walk a very fine line between implementing innovative techniques and moving too fast to the point of misunderstanding and revolt," says Yackey. "In working with various groups in the organization, we have learned to be flexible and tailor our approach to our audience. For instance, we found that generally speaking those in IS appreciated and were thirsty for structure, a clear methodology and documentation, while those on the clinical side feel they have plenty to do already without Corporate coming and adding to it." A key question in the portfolio selection process is therefore "Will this impact hospital staff?" To their credit, the EPMO has found that, each team that they have worked with has become a believer in project management and in the organizational changes they have initiated.

How the EPMO Has Changed Norton

Today, project management is embedded in Norton's culture. According to Yackey, "We have moved the organization along a continuum from chaos to structured chaos to planned structure and strategy."

Enterprise Portfolio Management. In 2008, the EPMO moved to the next level of enterprise strategic planning using portfolio management to limit and prioritize programs and projects based on urgency, budget and resource needs. Yackey stresses, "The business strategy is the starting point. By breaking down the overall strategy, we can organize the work to be done in a series of departmental portfolios which are then broken down into programs (if appropriate) and projects. Now everyone can see how they will personally help implement Norton Healthcare's strategic plan." In addition, compensation is linked to meeting objectives.

Norton Healthcare's EPMO provides a striking example of how project management can be tailored to industry and culture, while at the same time injecting that culture with a boost of efficiency and realism created by rational planning and timely delivery of projects.

2008: Accident Fund Insurance Innovation and Planning Department

It's No Accident: This Strategic PMO is One of a Kind

Accident Fund Insurance is a provider of workers' compensation and disability insurance—the 15th largest such company in the U.S. In 1999, a Project Management Office (PMO) was created within their Information Systems (IS) department, with a focus on managing IS projects. Its success led to the creation, five years ago, of a new entity and kicked off an organizational journey unique in the industry.

Innovation, Indeed. The new PMO was deliberately moved out of IS to be independent of any business unit in order to maintain focus and objectivity, and combined with the E-magine unit, which was focused on researching and implementing new e-business strategies. The combination of these two groups created a new department called Innovation and Planning (I&P). The I&P team is a hopper for business ideas from the executive team and from staff and customers. I&P owns these ideas right from initial research and exploration. The investments that are deemed worthy are prioritized for resources, pass through several decision gates, and are implemented as part of the enterprise strategic plan.

Since the time these two teams were combined, I&P has been directly accountable to the executive staff through the Vice President of Planning. This VP role is responsible for leading the creation and execution of the strategic plan for the enterprise. In addition to this direct channel to the strategic planning process, I&P was given its own department budget for staff, tools and more importantly the responsibility of these capital budget dollars that are associated with implementing these strategic projects. This, in turn, placed accountability and authority for these strategic projects specifically under the new department.

The broad portfolio of strategic business projects and programs that I&P has been responsible for includes:

- Large systems development projects
- Business expansion for all product lines
- Mergers and acquisitions projects
- R & D activities.

I&P has continued to mature, adapt, and evolve with the changing needs of a growing enterprise. Through these changes, there are additional PMO innovations that the 2008 award judges believe not only differentiate the I&P from other PMO's, but make it one of the best in class in *any* industry.

Some of these unique differentiators include:

- A clear linkage to and ownership of the five-year strategic plan through each annual planning process.
- The VP of Strategic Planning participates directly with CEO and senior staff and actively facilitates the ongoing dialogue on the strategic plan during monthly offsite meetings with all VPs and the CEO.
- The PMO takes priorities and direction from a cross-functional Executive Steering Committee (ESC). The PMO team institutionalized a dual project governance structure: the ESC for strategic initiatives and Project Prioritization & Management (PPM) Committee for tactical department project prioritization, and brought acquired subsidiary companies along into these processes.
- The I&P team directly facilitates these committees through scenario planning, prioritization, decision gates, cost-benefit analysis (CBA) reviews, resource allocation, and project sequencing decisions.
- The project managers "own" the initiatives right from concept exploration through closeout and benefit recognition—regardless of the type of project (IT, facilities, M & A, R & D, etc.).

- The I&P team is responsible for a bringing a broad range of methodologies to bear for executing projects (i.e. methodologies for Systems Development projects, Agile projects, R & D projects, Mergers & Acquisitions, enterprise change management, and facilities construction).
- The PMO is responsible for linking the annual plan (a subset of the five-year strategic plan) process directly to the annual budgeting process across the enterprise.
- The PMO manages project team roles and forecasts resources right down to name-level assignments through a shared resource pool across projects and operates with geographically remote staff who reside in the subsidiary companies.
- The PMO is responsible for the post-project "Benefits Realization Process" (Did we recognize the benefits outlined in the CBA if not, why not? If not yet, then when?)
- The PMO now has a sub-team (the Vendor Management Office, or VMO) within I&P that manages key enterprise vendor relationships, contract negotiations, RFPs, SLAs, vendor audits, and creative sourcing opportunities.
- Finally, the negotiated cost reductions and recoveries that the VMO team provided in 2007 *funded the entire I&P Department*—meaning that the department can likely be self-funded in the future, and laying to rest forever the thorny question of PMO value.

Because of the broad range of initiatives the team is responsible for, the staff must be able to understand not only the strategies and business objectives for the enterprise, but how those objectives can be met through various combinations of people, processes, and technology changes. In addition to focusing on benefits realization, the PMO institutionalized and owns the enterprise project governance structures—the ESC for strategic initiatives and the PPM for tactical project prioritization.

After winning the PMO of the Year Award in 2008, the Accident Fund PMO was featured on the cover of *PM Network* magazine in 2009.

Note: Case summaries about all past winners and finalists for the PMO of the Year Award are archived at http://www.pmsolutions.com/pmoaward/past-winners/

Appendix C: Project Management Assessment and Recommendation Report

(For COMPANY XYZ)

Background

Company XYZ is a service bureau for the processing of Mutual Fund services. The organization is the business definition and development department of Company XYZ. Their line of business (LOB) partners and corporate staff are located in Delaware.

Company XYZ has gone through a significant growth period over the past several years, expanding their business solutions and customer base. The target organization, for whom this report has been developed, had to support this business growth, which called for new development, many enhancements to existing systems, and adding several outside packages to their system mix. It also increased the size of the infrastructure in people and systems.

The senior vice president, (name), has recognized that there is a problem in the planning and management of projects within the department.

The continued business has brought many new employees on board. These employees differ significantly from the current population, in that they are more educated and have had other work experiences. Various long-time employees have a strong loyalty to Company XYZ and use their own loosely structured processes to get the job done.

PM Solutions was engaged by Company XYZ to address their project management methods, techniques, and practices by providing the following:

1. Assessment of current project management practices/methodologies
2. Summarization of assessment and recommendations for improvements
3. Presentation of a two-day Project Management (PM) Essentials course

The PM Essentials class was delivered successfully on May 10 and 11, 1999. This document represents the culmination of Steps 1 and 2 above.

The purpose of this assessment is to determine the areas of project management in which Company XYZ requires improvement, and to understand those areas in which Company XYZ has the greatest strengths, or "best practices."

The assessment was accomplished in two phases:

- Survey questionnaire
- Personal interviews

Survey Questionnaire

The PM HealthCheck™ survey or questionnaire is a self-assessment instrument designed to provide a means of determining the health and strength of the project management practices being applied in the organization and on projects. It consists of ten candid, introspective questions on each of the nine major knowledge areas covered in the Project Management Institute (PMI) *Guide to the Project Management Body of Knowledge* (*PMBOK*®). These areas include:

- Scope management
- Time management
- Human resource management
- Communication management
- Risk management

- Quality management
- Cost management
- Procurement management
- Project integration

The survey questionnaires were completed by 28 people, between April 16 and 23, 1999.

Personal Interviews

Seventeen people and the senior vice president were interviewed onsite at Company XYZ on May 4 through May 6, 1999. They included a random sampling of all employees who had completed the questionnaire and represented all segments of the department.

The interviews focused on individuals' background and experiences, organization strengths and weaknesses, validation/understanding of individuals' questionnaire results and issues, and what was needed for the individual to be successful.

The results of the surveys and interviews are provided in the following Summarization section. Nearly everyone interviewed was very candid and concise. There is a recognized need for improvements and that there are better project management techniques and practices, *but, most importantly, the attitude is generally one of openness and willingness to change*. This attitude is a critical success factor that is pivotal to the success of project management improvements.

Summary

This section summarizes the findings of the HealthCheck survey and the subsequent interviews, and provides recommendations.

This report is brief and to the point. Supporting detail has not been documented in this report due to time and budget constraints.

In *The Chaos Report* by Standish, it is reported that only 16 percent of IT projects are successful. The reasons these 16 percent are a success is typically:

- Clear requirements
- User involvement
- Executive management support

- Proper planning
- Realistic expectations

The above research is noteworthy, as the lack of the listed attributes above are some of the reasons for the issues at Company XYZ.

- Clear/concise requirements are the exception.
- User involvement is not consistent and requirements often change throughout a project.
- There is no incentive for developing consistent, repeatable processes that will ensure effectiveness, efficiency, and user delight.
- The business analysts establish expectations with little regard to input from systems and programming.

It is important to note that there are pockets of effective planning, primarily from new personnel who have used consistent project management methodologies.

On both the HealthCheck survey and the interview sessions, Company XYZ personnel confirmed senior management's belief that there are *critical* concerns in the area of project management. Perhaps the clearest portrayal of the nature and gravity of the problems can be seen in the chart shown below, which summarizes the survey results. Note: A complete set of charts from the HealthCheck survey and a copy of the actual survey are contained in the appendices.

Each person's questionnaire results were grouped by "job category" (CIO, business analysts, and team leaders) or business unit (FSA, accounting, and transfer). The scores from all categories were used to find an overall average, which is shown on the following charts as 1.6 (Figure C.1 to Figure C.3). In terms of meaning, this translates loosely to "Intensive Care" to "Critical" when plotted against the PM Solutions Maturity Model. In this writer's opinion, this situation demands immediate attention.

The noticeable "gap" between team leaders and the balance of the organization (below) is notable, as it is representative of poor communications processes and/or a lack of clear understanding of roles and responsibilities.

The interview process supported the findings of the survey.

There were a number of areas perceived by the interviewees as strengths and are recognized as a credit to the management team.

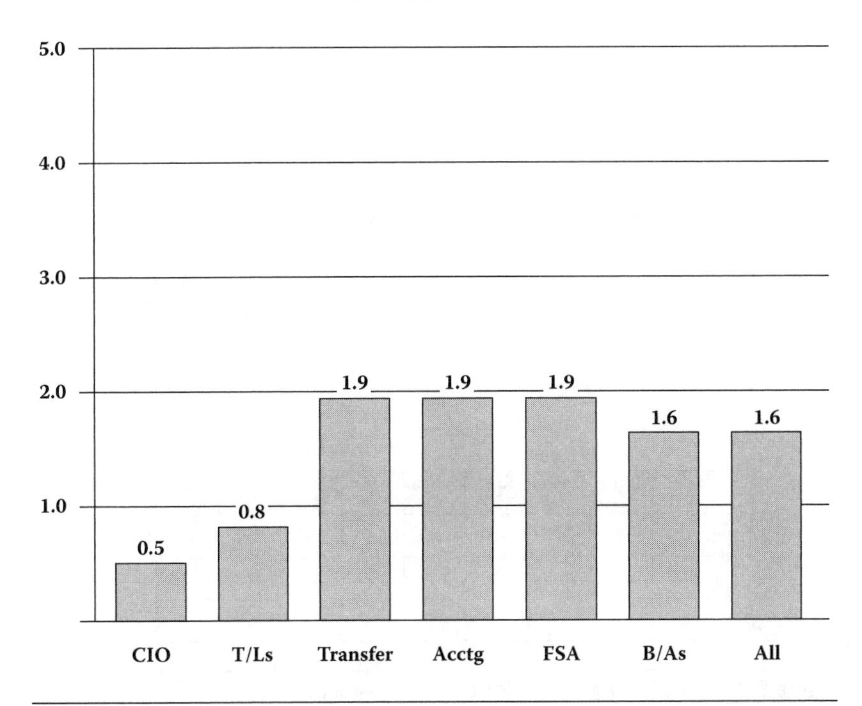

Figure C.1 PM Health Check summary by grouping.

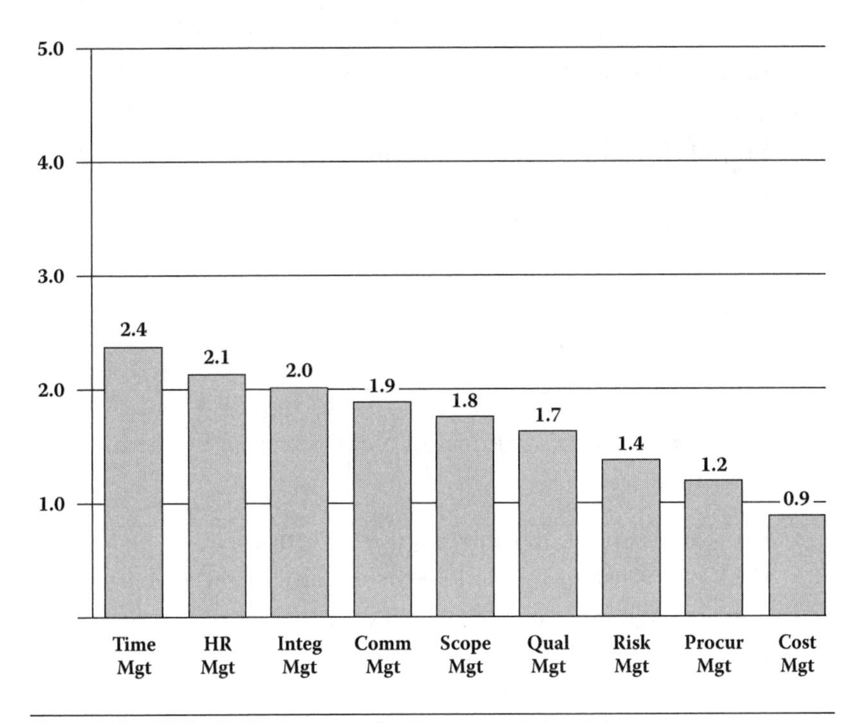

Figure C.2 PM Health Check summary for all responses, by knowledge area.

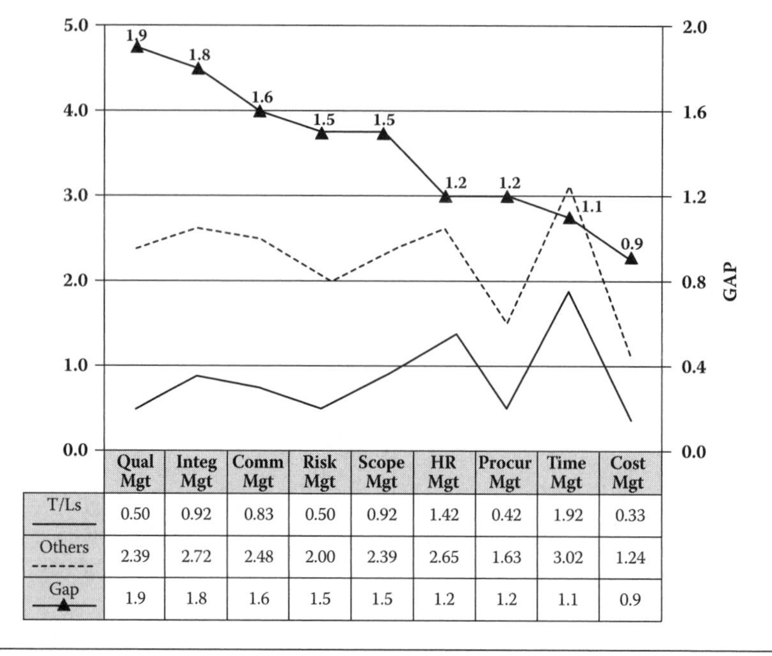

	Qual Mgt	Integ Mgt	Comm Mgt	Risk Mgt	Scope Mgt	HR Mgt	Procur Mgt	Time Mgt	Cost Mgt
T/Ls ———	0.50	0.92	0.83	0.50	0.92	1.42	0.42	1.92	0.33
Others -------	2.39	2.72	2.48	2.00	2.39	2.65	1.63	3.02	1.24
Gap —▲—	1.9	1.8	1.6	1.5	1.5	1.2	1.2	1.1	0.9

Figure C.3 Health Check gap: Team leaders (T/L) vs. All others.

- Good people with a lot of pride
- Business knowledge
- Technical skills
- Projects (usually) completed on time
- Loyalty
- Commitment to get things done
- Customer orientation (customer accommodation)
- Variety of work
- Do a lot with little
- Low turnover

Likewise, there were a number of areas perceived by the interviewees as weaknesses:

- Poor communication (horizontal and vertical)
- Lack of consistent methodology/standardization for business operations (not organized to be project efficient)
- No consistent approach for requirements definition (Note: Effort is currently underway to develop a tool/template to address this.)

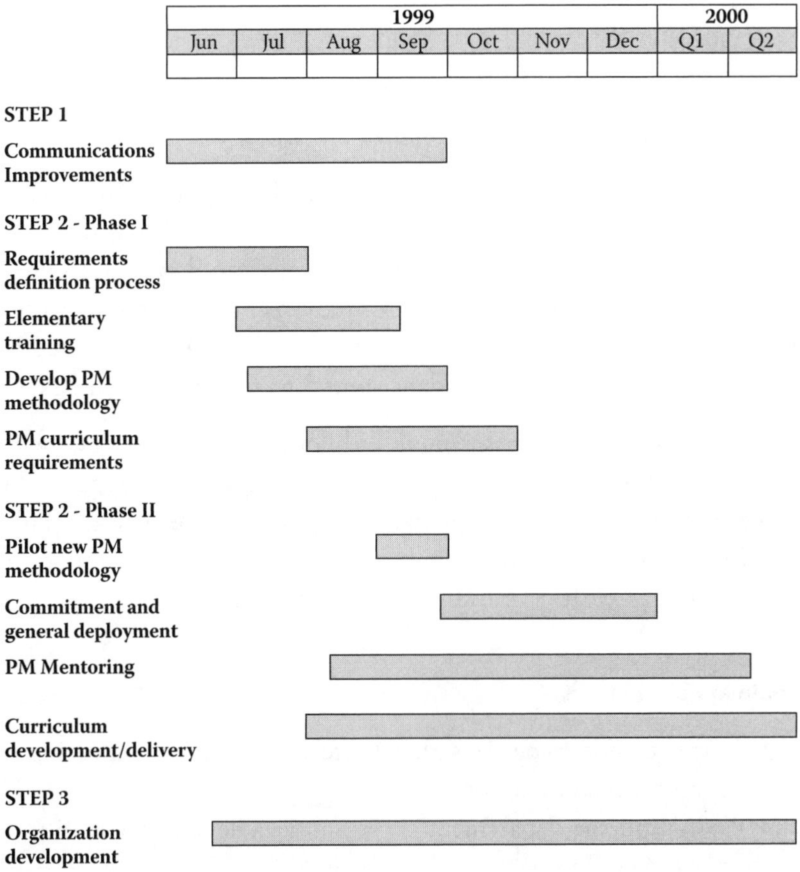

	1999							2000	
	Jun	Jul	Aug	Sep	Oct	Nov	Dec	Q1	Q2

STEP 1

Communications Improvements

STEP 2 - Phase I

Requirements definition process

Elementary training

Develop PM methodology

PM curriculum requirements

STEP 2 - Phase II

Pilot new PM methodology

Commitment and general deployment

PM Mentoring

Curriculum development/delivery

STEP 3

Organization development

Figure C.4 Organizational development schedule.

- Business unit's relationship with systems and programming
- Lack of historical project data
- Lack of enterprise resource information/allocation/management (work prioritization, competition for resources, assignment of workloads, etc.)
- Lack of effective tools/techniques (e.g., estimating)
- Poor or nonexistent change control
- No good tool for project status or summarization across the organization
- Lack of risk analysis/management
- Unclear roles and responsibilities, accountabilities, and performance requirements

- Lack of consistent QA processes and practices
- Poor project management expertise/knowledge
- Recognition by headquarters, isolation
- Design and architecture of current applications

Given the fact that many people mentioned the same areas, we conclude that most people in the organization understand that there are better ways in which to plan and manage projects and to conduct business. They appear to be looking for improvements in communication, a set of common practices that are institutionalized and supported by management in addition to guidance and direction on deployment.

Note: The scope of the assessment did not allow for interviews with "users and/or customers." As a result, this is a gap in this report. The customers and users of the organization's processes are critical suppliers of key inputs and their perception of problems/issues is important in addition to a clear definition of their expectations.

Summary of Findings

- Management recognizes that there is a need for improving organizational capabilities.
- Loyalty to the department is very noticeable and is a significant reason for operational successes.
- The organization is populated with many good people who take pride in getting things done with little, in the face of formidable odds.
- A majority of the organization recognizes the need for change and appears to be ready to embrace it.
- Communication or lack thereof, both horizontal and vertical, is a major issue.
- There are no consistent, repeatable and/or accepted processes utilized within the organization.
- Crisp, committed requirements definition for projects, signed off by customers, is an exception.
- New employees have experienced organized projects in their previous employment and are disillusioned by the lack of organization and standards.

- Many of the current employees, particularly the lead programmers, programmer analysts, and team leaders, have been in the organization for 10 to 25 years. They have little or no standard against which to base their current job.
- New employees brought into management positions face difficulty in being accepted into the organization.
- While the term *team* was used on occasion, there is no evidence of effective teaming occurring.
- There is a strong desire to improve the current environment, but little motivation to "take the lead." People do not feel empowered.
- Each S&P team leader manages his/her group with varied levels of skills.
- People are not motivated to work to standards.
- Job descriptions exist, but serve little value and appear to be used only for pay banding.
- A systems life cycle (SLC) process exists and was found to contain processes that could be modified to meet the department's needs, but is not used and numerous interviewees were not aware of its existence.
- There are no clear and consistent definitions of roles and responsibilities. This has resulted in some people carving out the role they want to fill within the organization.
- Financial accountability is not defined at the project level nor is there clearly defined success measures that performance (project team or individual) can be measured against.
- People are not recognized for their hard work and meeting deadlines. They are expected to do what is necessary to meet deadlines, which have been imposed from outside their group.
- Systems and programming has little say in setting deadlines, estimating the time necessary to get work done, or defining the work that needs to be done.
- Cross training or movement within the organization does not occur. There is little structured emphasis on career development.
- Successes are not celebrated or shared openly nor are there department activities to reenforce teaming and bonding.

- There is a team working on a new requirement documentation tool, but little knowledge this is occurring. The tool needs to be surrounded by a process.
- Many of the department's successes are based on personal relationships and crisis management.
- There are no consistent performance reports or an "Executive Dashboard" with key performance metrics.
- Relationships between the department and the lines of business (LOBs) are generally weak. The perception is the LOBs do not have a high regard for the department.
- The organization is not structured effectively for efficient execution of business requirements.

Recommendations

The four primary areas that need immediate care:

- Communication
- Methodologies/processes
- Organization development/department operations
- Training

The following are recommendations to improve the department's performance and are all deemed vital.

Establish Communications with and within the Department

- Develop a communication plan for the organization.
- Conduct department meetings monthly. Introduce new employees, recognize successes, and discuss new business.
- Conduct LOB meetings on a regular basis to discuss issues, resolve problems, establish procedures, update, etc.
- Establish a simple newsletter to be distributed on a periodic basis.
- Have department get-together periodically, outside of the work environment, to celebrate successes, reinforce worth, and encourage bonding.

Improve Methodologies/Processes

- Establish a clear definition of what a project is and is not.
- Develop and deploy a standard *project management methodology* tailored to the department's needs. It should contain the project management processes and also templates for the major deliverables from the project management process. It should be made clear within this guide how small projects should be handled (are there certain steps that can be left out, etc.) It should be made clear to all, both within the department and in the client areas, that this methodology is the cornerstone for success. Ensure the first edition is simple to use so people are not overwhelmed or use it as an excuse for "getting in the way of getting the job done."
- Develop a process for requirements definition immediately.
- Devise and institute a change management process that must be utilized on every project.
- Devise and institute a risk management process that must be utilized on every project.
- Company XYZ needs a portfolio management process. The organization should create a detailed, documented process for managing the project portfolio. We recommend that the responsibility for managing the portfolio rest with a Project Office. This includes managing project requests, prioritization, etc.
- Company XYZ has a systems development life cycle standard in house, although it has not been institutionalized and used. Company XYZ should tailor and adopt this as a standard operating practice.
- Develop a sign-off process at project closure (sign-off by LOB and operations) and ensure that experiences and lessons learned are shared.

Organization Development/Department Operations

- Establish a standards department, with responsibility as a service organization to the project and support teams, which provides expert knowledge, advice and assistance in quality management, project management, and process management.

- Project prioritization
- Leadership and appropriate structure to improve individual project performance
- Departmentwide resource management, resource procurement, and project deployment
- Work-plan setup, review, monitoring
- Standardized systems development methods and frameworks
- Risk management across the portfolio
- Knowledge management and sharing of best practices
- Consistent project management practices throughout the department
- Increased number and capacity of project managers
- Portfolio view of major projects
- Coordination of process change/continuous improvement
- Coordination of departmentwide planning and budgets

- Formally assign empowered project managers and establish budgets for all projects.
- Establish a project and service call prioritization process and dedicate resources to a help desk function to handle service/support calls exclusively and support them with the service call/support process.
- Consider modifying the organization structure to place appropriate resources, supporting a specific line of business, into the same functional unit.

—— or ——

- Form project teams at the inception of a project, assign an empowered project manager, and place control of the resources under the project manager until the project is closed.
- Establish clear and consistent roles and responsibilities. Establish standards of performance. Establish clear accountabilities and performance objectives for all positions that support the overall goals and objectives of the organization. Reward accordingly. Ensure that career development

objectives are defined and processes are implemented to encourage professional development.

- Establish a recognition program. Allocate discretionary dollars to managers/team leaders to recognize their employees. Teach managers how to recognize people. Manage by wandering around. Know what people are doing and how they feel. Measure the satisfaction of all associates on a regular basis.
- Establish a management infrastructure, annual operating budgets, and other management requirements to appropriate levels to promote accountability.
- Develop a business plan and recommendation for charging back all work to the LOBs. Perceived value will improve as well as efficiency. The current practices reenforce bad business behavior.
- Reconsider the plan to move quality assurance out of the organization. The quality of the deliverable and all the testing should be the responsibility of the project teams and the project manager.

Training

- Company XYZ has conducted training in PM Essentials. The learning from this class must be put to practical use throughout the department.
- The business clients need to be trained in the methods and value of project management. It is our suggestion that they be provided a PM Essentials overview to promote interaction and understanding between the department and business client personnel.
- Company XYZ should conduct training in their project planning/control tool (MS Project 98). This will reinforce knowledge already acquired by some of the personnel and provide critical skills for others.
- Company XYZ should consider an advanced project management training course after the *Project Management Handbook* is deployed and formal assignment of project managers is implemented.
- Establish a comprehensive orientation program for new employees.

Recognizing that funding for all improvements immediately is not practical, we recommend a prioritization list be developed and that the list becomes the basis for the milestones of a project plan designed to improve the efficiency and client satisfaction of the department. A shared objective also should be established, across the management team, to implement (develop and deploy) the milestones of the project, on schedule.

Note that PM Solutions has the capability of performing many of the functions in the above list of recommendations. Examples include:

- Creation or tailoring of a project management methodology
- Training on project management (single courses or an entire curriculum leading to a Masters Certificate in Project Management)
- Training on a variety of project management tools
- Project manager mentoring
- Integration of PM processes with an SDLC
- Project Office tailoring and mentoring

Recommended Action Plan

Step I: Communication

- Implement "quick fixes" immediately; e.g., schedule and execute monthly/quarterly department meetings and LOB team meetings, with crisp agendas. (ASAP)
- Establish facilitated sessions with small groups of associates to "listen" for requirements and solicit input. (ASAP)
- Establish a cross-functional team, chaired by HR or a neutral facilitator. The mission and objective of the team is to develop and deploy a comprehensive communication plan for the organization and supporting processes by October 1. (July to September 1999)
- Develop an orientation seminar/tool for new employees.

Step II: Project Management Training, Methodologies/Processes

Phase I

- Develop a crisp, user-friendly requirements definition process immediately, working closely with key stakeholders, and implement throughout the organization. (June to July 1999)

- Provide PM Essentials course to the balance of the organization and selected LOB partners. (July 1999)
- Provide MS Project 98 course to all project managers as well as team members who have a need-to-know. (July to August 1999)
- Develop a standard project management methodology and a process guide (*Project Management Handbook*) tailored to the department's needs. (July to September 1999)
- Develop curriculum requirements for project management training and establish a schedule to satisfy requirements. (August to October 1999)

Phase II

- Pilot and deploy the Company XYZ project management methodology.
  - Pilot the process on two projects. (September 1999)
  - Gain commitment from all teams. (September to October 1999)
  - Deploy the new process throughout the organization and measure compliance. (October to December 1999)
- Provide project management mentoring to select project teams to enhance knowledge and skill transfer. (September 1999 to Q2 2000)
- Build or contract for courseware to satisfy the PM curriculum requirements defined in Phase I. (October to December 1999)
- Implement the PM curriculum throughout the organization and celebrate/reward levels attainment. (Year 2000)

Step III: Organization Development/Department Operations

- Establish a standards department with responsibility as a service organization to the project and support teams, to provide expert knowledge, advice, and assistance in project, process, and quality management. (ASAP)
- Establish clear and consistent roles and responsibilities. Establish standards of performance. Establish clear accountabilities and

performance objectives for all positions, that support the overall goals and objectives of the organization. Reward accordingly. Ensure that career development objectives are defined and processes are implemented to encourage professional development. (Q3 1999)
- Establish a management infrastructure, annual operating budgets, and other management requirements at appropriate levels, to promote accountability. (Q4 1999)

The PM Solutions Assessment Team wishes to thank (names), and those associates in the organization who gave their valuable time to complete the survey questionnaires and participate in the interviews. The frankness and understanding of the issues is very encouraging.

Index

D